POST PANDEMIC L2 PEDAGOGY

PROCEEDINGS OF THE LANGUAGE TEACHER AND TRAINING EDUCATION VIRTUAL INTERNATIONAL CONFERENCE (LTTE 2020), 22–25 SEPTEMBER, 2020

Post Pandemic L2 Pedagogy

Edited by

Kristian Adi Putra
Prince Sattam Bin Abdulaziz University, Saudi Arabia
Universitas Sebelas Maret, Indonesia

Nur Arifah Drajati
Universitas Sebelas Maret, Indonesia

Routledge is an imprint of the Taylor & Francis Group, an informa business

Typeset by MPS Limited, Chennai, India

Library of Congress Cataloging-in-Publication Data
Applied for

Published by: Routledge
Schipholweg 107C, 2316 XC Leiden, The Netherlands
e-mail: Pub.NL@taylorandfrancis.com
www.routledge.com – www.taylorandfrancis.com

ISBN: 978-1-032-05807-8 (Hbk)
ISBN: 978-1-032-05809-2 (Pbk)
ISBN: 978-1-003-19926-7 (eBook)
DOI: 10.1201/9781003199267

Post Pandemic L2 Pedagogy – Adi Putra & Arifah Drajati (Eds)
© 2021 Taylor & Francis Group, London, ISBN 978-1-032-05807-8

Table of contents

Post Pandemic L2 Pedagogy – Adi Putra & Arifah Drajati (Eds)
© 2021 Taylor & Francis Group, London, ISBN 978-1-032-05807-8

Scientific Committee

Post Pandemic L2 Pedagogy – Adi Putra & Arifah Drajati (Eds)
© 2021 Taylor & Francis Group, London, ISBN 978-1-032-05807-8

Committees

Patron
Jamal Wiwoho
Rector of Universitas Sebelas Maret (UNS)

Mardiyana
Dean of the College of Education, UNS

Steering Committee
Joko Nurkamto
Head of the Master's Degree Program in English Language Education, UNS

Organizing Committee
Nur Arifah Drajati
Chair

Kristian Adi Putra
Secretary

Sri Haryati
Treasurer

Sumardi
Coordinator of Event

Anis Handayani, Mifta Muriska Isya & Fatimah Zahro Assidiqoh
Members

Lita Liviani Taopan
Coordinator of Public Relations

Niken Sri Noviandari & Ma'rifatul Ilmi
Members

Adi Irma Suryadi
Coordinator of IT Support

Surya Agung Wijaya & Rudy Setiyanto
Members

Rizka Junhita Rahma Wanodya
Coordinator of Documentation

Agustina Tyarakanita & Lusiana Pratiwi
Members

Hilda Rakerda, Alfina Fadilatul Mabruroh & Reni Puspitasari Dwi Lestariyana
Coordinator of Publication

Post Pandemic L2 Pedagogy – Adi Putra & Arifah Drajati (Eds)
© 2021 Taylor & Francis Group, London, ISBN 978-1-032-05807-8

Acknowledgements

Language Teacher and Training Education (LTTE) 2020 Virtual International Conference would not have been possible without the financial support from the Institute for Research and Community Service (LPPM), Universitas Sebelas Maret (UNS). We are especially indebted to the Rector of UNS, Prof. Dr. Jamal Wiwoho, M.H., the Dean of the Faculty of Teacher Training and Education of UNS, Dr. Mardiyana, M.Pd., and the Head of the Master's Degree Program in English Language Education of UNS, Prof. Joko Nurkamto, M.Pd., whose advice, encouragement, patience, and enthusiasm have made the success of this conference possible.

We also would like to thank all the invited keynote speakers, Dr. Joko Nurkamto (Universitas Sebelas Maret, Indonesia), Dr. Utami Widyati (Universitas Negeri Malang, Indonesia), Dr. Ju Seong Lee (The Education University of Hong Kong), Dr. Jonathon Reinhardt (The University of Arizona, USA), Dr. Yilin Sun (Seattle Colleges), Dr. Christine Coombe (Dubai Men's College, UAE), Dr. Jonathan Newton (Victoria Wellington University, New Zealand), Dr. Chan Narith Keuk (Royal University of Phnom Penh, Cambodia), and Dr. Alice Chik (Macquarie University, Australia), for their valuable knowledge sharing with all the participants of this conference.

We are also extremely thankful for all the committees of this conference who have tirelessly and patiently helped us from the beginning until the end. We would like to thank our colleagues, particularly Sri Haryati, S.Pd., M.Pd., Dr. Sumardi, M.Pd., for their continuous and endless support along the way. We are sincerely grateful for the help from a lot of our current and former students, including Lita Liviani Taopan, Anis Handayani, Mifta Muriska Isya, Fatimah Zahro Assidiqoh, Niken Sri Noviandari, Ma'rifatul Ilmi, Adi Irma Suryadi, Surya Agung Wijaya, Rudy Setiyanto, Rizka Junhita Rahma Wanodya, Agustina Tyarakanita, Lusiana Pratiwi, Hilda Rakerda, Alfina Fadilatul Mabruroh, and Reni Puspitasari Dwi Lestariyana. We owe them for their dedication to make this conference successful.

Last but not least, we also would like to thank our colleagues who have been willing to be the reviewers for this proceeding and all the editorial teams, particularly Lita Liviani Taopan, who have worked days and nights to help us regardless of the different time zone and her busy schedule. We have been extremely blessed with endless support from them.

Keynote speeches

Post Pandemic L2 Pedagogy – Adi Putra & Arifah Drajati (Eds)
© 2021 Taylor & Francis Group, London, ISBN 978-1-032-05807-8

Teachers' roles in fostering EFL learner autonomy: A literature review

Joko Nurkamto
Universitas Sebelas Maret, Indonesia

ABSTRACT: Over the last three decades, there has been a radical shift in the direction of English Language Teaching (ELT) from a teacher-centered point of view to a learner-centered perspective (Teng 2019). This shift in priority has suggested the need for a reorientation of teachers' roles to share the responsibility with the students and to provide them with more opportunities to take charge of their own learning (Benson 2011). This has subsequently led to a significant increase in attention on Learner Autonomy (LA) in ELT. As an educational aim of ELT, LA has attracted the interest of many ELT scholars. However, few discussions have been specifically devoted to the efforts made by teachers to develop LA. This paper aims to explore the role of teachers in fostering LA. By understanding this issue, EFL teachers should be able to maximize the role they play in developing their students' autonomous learning.

Keywords: English language teaching, learner autonomy, sharing the responsibility, teachers' roles

1 INTRODUCTION

Over the past thirty years, we have witnessed a sharp shift in the direction of learning, from a teacher-centered emphasis to a learner-centered focus (Teng 2019). This change in the swing of the pendulum has had implications on the need for a reorientation of the teacher's role, from a person who simply transfers knowledge to students or learners, to become a facilitator who helps learners construct their own knowledge (Richards & Rodgers 2014; Westwood 2008). This is based on the premise that knowledge is not a pre-existing entity, ready to be passed on to the learner, but rather something that must be constructed by the students themselves as the ones who are learning (Shieber 2019). Therefore, in teaching and learning activities, a more important question to ask is whether the students have learned well in such a way that they have acquired new knowledge.

In this context, there are at least three tasks that must be performed by the teacher. The first task is to guarantee the students' understanding of the material they are studying (cognitive level) (Wierzbicka 2011), for example by checking their understanding (Fisher & Frey 2007). The second task is to help students develop their own metacognitive awareness, which they can use to monitor and manage their thought process in an effort to gain knowledge (Ku & Ho 2010; Muijs & Reynolds 2018). The third task is to develop students' learning autonomy (LA) – the capacity to manage their own learning, both independently and in collaboration with others, both inside and outside the classroom (Benson 2011; Benson 2016; O'leary 2014). Developing LA is the ultimate goal of a teacher (Teng 2019).

There are numerous existing articles that discuss LA but very few which elaborate specifically on the teacher's role in developing LA, especially in the field of ELT. Therefore, in this paper, I wish to explore the above topic, based on a review of relevant literature.

2 DESCRIPTION OF LEARNER AUTONOMY

Several different terms are connected to LA, namely self-instruction, self-regulation, independent learning, self-access learning, and self-directed learning (Teng 2019). These terms, which are usually used interchangeably, refer to the same meaning, that is, a capacity that learners possess and perform to manage their own learning (Benson 2011; Oleary 2014; Teng 2019). The word "manage", in this context, refers to the learners' ability to successfully organize their learning that covers three dimensions, namely learning management, cognitive process, and learning content (Benson 2011). Learning management deals with the cognitive competencies that underlie the act of planning, enacting, assessing, and evaluating learning. Cognitive processing refers to the mental activities to monitor and control learning processes. Learning content is concerned with language learning materials that learners select, learn, and practice (Benson 2011).

Learners who are autonomous can determine their learning aims and objectives, define language contents, select methods, and techniques, monitor learning progress, assess learning achievement, and evaluate what has been done and acquired (Teng 2019). In addition, autonomous learners are disciplined, reflective, curious, flexible, interdependent, responsible, creative, confident, independent, skilled in retrieving information, skilled in learning, and able to develop and use criteria (Benson 2011).

3 FOSTERING LA

Benson (2011) identifies six approaches of LA development, namely a resource-based approach, technology-based approach, learner-based approach, classroom-based approach, curriculum-based approach, and teacher-based approach. Resource-based LA development emphasizes the availability of relevant learning resources (such as self-access centers, libraries, e-books, and e-journals) and provides opportunities for learners to utilize these learning resources in accordance with their own needs. Technology-based LA development places emphasis on the availability of technology, both on and off campus, and encourages students to make use of the available technology to maximize their learning, which includes access to existing learning resources. Learner-based LA development prioritizes the development of skills needed in order to utilize existing learning resources and technology. In this context, the development of student capacity is extremely important.

Classroom-based LA development focuses on students' involvement in the decision-making process in relation to their own daily learning management, such as planning and learning assessment. This requires training, and its implementation is in stages (at first under supervision, and then independently). Curriculum-based LA development places emphasis on students' involvement in the process of curriculum development, beginning with a needs analysis and continuing with goal setting, syllabus development, learning materials development, learning methods, and learning assessment. In this case, negotiation between the lecturer and the students is essential. In addition, it is highly beneficial for lecturers to provide scaffolds. Teacher-based LA development places importance on the roles of the teacher or lecturer (as a facilitator, motivator, counselor, and resource manager) in the implementation of autonomous learning development through the various approaches outlined above. In order to carry out these roles, lecturers themselves must also become autonomous teachers. In connection with this, the involvement of teachers/lecturers in ongoing teacher/lecturer development programs is crucial.

4 TEACHERS' ROLE IN FOSTERING LA

Teachers have an important role to play in the development of LA. Teachers should not only wait and encourage their students to become autonomous learners but should also prepare them to be active and independent learners (Teng 2019). A similar idea is proposed by Tran and Duong (2018), who states that one of the main responsibilities of teachers is to help learners to understand how to become autonomous learners because they cannot become effective autonomous learners on their own. I believe that fostering LA requires more than telling learners to become autonomous learners and then simply hoping for the best. Instead, learning autonomy should be taught explicitly, directly, and patiently. In this section, I will describe a number of roles that teachers can play in developing LA.

Tran and Duong (2018) identify four roles a teacher can play in developing LA, including the role of a facilitator, a counselor, a resource manager, and an assessor. As a facilitator, the teacher can show learners how to manage their own learning, for example by setting feasible learning aims and objectives, defining relevant learning materials, selecting appropriate learning methods, assessing their learning, and evaluating what they have done and acquired. As a counselor, the teacher can offer the learners advice on effective ways to achieve their learning aims and objectives efficiently. As a resource manager, the teacher can provide learners with a variety of academic information that they may need to accomplish their work effectively. As an assessor, the teacher can give learners feedback and advice on how to improve their work. The learners must expect that their work is assessed by a teacher who has solid knowledge and advanced skills in language assessment.

With regard to the teacher's role as a facilitator, Benson (2011) proposes technical and psycho-social support in the development of LA. The key features of technical support are as follows: (1) helping students to plan and implement independent learning, by carrying out a needs analysis, setting study goals, and selecting relevant learning material; (2) helping students to carry out self-evaluation, which includes assessing initial language proficiency, monitoring learning progress, and performing peer and self-assessment; (3) helping students to develop metacognitive awareness to support the two aforementioned items, for example by training students to develop the skills needed to identify appropriate learning styles and learning strategies.

The key features of psycho-social support include (1) personal qualities of the facilitator, such as being attentive, supportive, patient, tolerant, empathetic, and open; (2) the ability to motivate learners, by strengthening their commitment, reducing or eliminating uncertainty, helping them to solve problems, and being willing to engage in dialogue; (3) the ability to promote the awareness of students to become autonomous learners.

Benson (2016) also offers suggestions for teaching strategies that can foster LA. These include: (1) preparing lessons that enable students to understand what they are going to learn so that the students will be actively involved in the learning; (2) using authentic learning materials and authentic language to increase the level of relevance of students' needs with the real world; (3) motivating students to carry out an independent inquiry as a means to construct knowledge; (4) involving students in instructional design, such as selecting learning material that matches their interests; (5) encouraging students to carry out dialogical interaction so that good interpersonal relationships can be established; and (6) prompting students to reflect on what they have learned, to identify the strengths and weaknesses of the learning implementation as a foundation for designing better learning.

In his research, Wang (2016) states that 88.6% of the teachers studied felt that they were developing LA. The methods used to foster LA were as follows: (1) asking students to carry out research about specific lesson topics; (2) asking students to evaluate and correct their own work; (3) using various forms of group work; (4) asking students to perform peer teaching; (5) giving online assignments; (6) asking students to read references outside the class and present their findings to their fellow students in the classroom; (7) organizing extracurricular activities, such as debates, speaking, and short plays; (8) giving students the opportunity to share their learning experiences with each other; (9) providing useful web sources and encouraging students to carry out independent learning; (10) discussing the importance of LA with students; and (11) offering the students training about independent learning strategies.

After reading seven references dating from 1975 to 2004, Han (2014) identified a number of roles that teachers play in developing LA. These include the teacher playing the role of (1) facilitator, assistant, and consultant; (2) a manager who creates a supportive and stimulating learning environment; (3) an active participant who monitors the course of the lesson, and a consultant who offers guidance about good learning methods; (4) a motivator who makes students aware of the importance of being responsible for their own learning; (5) a facilitator who initiates and supports the process of decision making, a counselor who responds to students' needs, and a learning resource who shares knowledge and expertise with the students; (6) a person who develops learning strategies to be used by the students; and (7) a person who builds assurance and increases students' confidence in LA.

5 CONCLUSION

In closing, I would like to stress that with the shift in focus of learning from "teaching" to "learning", teachers need to make students aware that they have a greater responsibility to learn. Students should no longer simply receive or accept "knowledge" from the teacher since knowledge is not a pre-existing entity that they can receive from another person. Instead, they must construct their own knowledge using methods that suit the needs and the contexts of the time and place where they are learning. In other words, they must become autonomous learners, or learners who have the capacity to manage their own learning, both independently and in collaboration with others, both inside and outside the classroom.

The teacher's responsibility in this context is to help and motivate learners to understand how to become autonomous learners because they cannot become autonomous learners by themselves. Some of the important roles a teacher can play include acting as a facilitator, a counselor, a learning resource, an assessor, an active participant, a learning manager, a learning partner, a trainer, and a motivator. Teachers need to build the assurance and increase the confidence of their students so that they can become autonomous learners.

REFERENCES

Benson, P. (2011). *Teaching and researching autonomy*. London, UK: Routledge.

Benson, P. (2016). Language learner autonomy: Exploring teachers' perspectives on theory and practice. In R. Barnard & J. Li (Eds.), *Language learner autonomy: Teachers' beliefs and practices in Asian contexts* (pp. xxxiii–xliii). Phnom Penh, Cambodia: LEiA.

Fisher, D., & Frey, N. (2007). *Checking for understanding: Formative assessment techniques for your classroom*. Alexandria, VA: ASCD.

Han, L. (2014). Teacher's role in developing learner autonomy: A literature review. *International Journal of English Language Teaching* 1(2), 21–27.

Ku, K.Y.L. & Ho, I. (2010). Metacognitive strategies that enhance critical thinking. Metacognition Learning, 5: 251–267.

Muijs, D & Reynolds, D. (2018). *Effective teaching: Evidence and practice*. Los Angeles, CA: SAGE.

O'leary, C. (2014). Developing autonomous language learners in HE: A social constructivist Perspective. In G. Murray (ed.), *Social dimensions of autonomy in language teaching*, pp. 15–36.

Richards, J.C., & Rodgers, T.S. (2014). *Approaches and methods in language teaching (third edition)*. Cambridge, UK: Cambridge University Press.

Shieber, J.H. (2019). *Theories of knowledge: How to think about what you know*. Virginia: The Great Courses.

Teng, F. (2019). *Autonomy, agency, and identity in teaching and learning English as a foreign language*. Singapore: Springer

Tran, T.Q. & Duong, T.M. (2018). EFL learners' perceptions of factors influencing learner autonomy development. *Kasetsart Journal of Social Sciences*, 1–6.

Wang, Y. (2016). Developing learner autonomy: Chinese university EFL teachers' perceptions and practices. In R. Barnard & J. Li (Eds.), *Language learner autonomy: Teachers' beliefs and practices in Asian context* (pp. 23–42). Phnom Penh: LEiA.

Westwood, P. (2008). *What teachers need to know about teaching methods*. Victoria: Acer Press.

Wierzbicka, A. (2011). Bilingualism and cognition: The perspective from semantics. In Vivian Cook and Benedetta Bassetti (eds.), *Language and bilingual cognition*, pp. 191–217.

Post Pandemic L2 Pedagogy – Adi Putra & Arifah Drajati (Eds)
© 2021 Taylor & Francis Group, London, ISBN 978-1-032-05807-8

The use of mobile learning applications in listening classes among Indonesian EFL students across gender

Delsa Miranty & Utami Widiati
Universitas Negeri Malang, Indonesia

ABSTRACT: Mobile learning takes the form of applications that can be accessed from the internet, making it possible for students to have more access to listening materials. However, research about mobile learning applications in listening courses at the university level is still limited. This study aims to identify EFL students' perceptions and preferences regarding mobile learning applications in listening classes across gender. Open-ended questionnaires were distributed through Google Forms to 120 ELT students in a public university in Indonesia to gain relevant information. The results showed that the male students tended to use more than one mobile learning application than the female students based on their preferences of video-based language learning as a learning tool. The study also revealed that the male and female students positively perceived the use of mobile learning applications in listening classes.

Keywords: EFL listening classes, gender, mobile learning applications

1 INTRODUCTION

In recent years, mobile learning has become an essential part of our lives as a language learning tool to realize educational goals. Gafni et al. (2017) state that mobile learning successfully enhances students' academic performance as it increases their ability to develop foreign language skills. Related to mobile learning applications, Jumaat et al. (2018) conducted a study on the students' preferences of mobile learning applications related to learning materials based on their needs and on the perceived best tool to provide an efficient learning method.

Research studies have suggested that mobile learning applications have helped students learn English in the listening class; mobile applications for English learning are useful tools designed for language learners, especially in listening through applications, than other language skills (e.g., Gangaiamaran & Pasupathi 2017). Jia and Hew (2019) found that students who applied mobile learning applications improved listening skills and performed significantly better than those who did not use mobile learning applications. In addition, previous studies have also explored mobile learning applications in listening classes among female and male students. Bembenutty and White (2013) found that females used self-regulation, more significant time, and self-control more efficiently than males. Some other studies indicate that female students spent more time on mobile learning than male students (Reychav & McHaney 2017; Viberg & Grönlund 2013). Unlike those findings, another study from Liu and Guo (2017) found that male students favored status and value orientation, but female students preferred social orientation. Cai et al. (2017) found that male students have better knowledge of the technology used than female students. In other words, although both males and females use technology, there seems to be gender differences in the use of mobile learning applications. Research studies revealed different results in how male and female students use the internet and other learning technologies.

Previous research has recorded some inconclusive findings related to students' preferences for mobile learning applications across gender in the listening class. Students' reasons for using mobile learning applications across gender among EFL students in the Indonesian context were rarely

DOI 10.1201/9781003199267-2

explored. Among the few studies, Darsih and Asikin (2020) investigated the English mobile applications that were considered useful and supported both for male and female students. The students chose mobile learning applications based on their preferences for the listening class. This study showed that the more mobile learning application downloaded and used, the better information they got to help their learning process. A questionnaire about mobile learning in Darsih and Asikin was adopted from Davis (1993) and modified according to the research context.

Based on the aforementioned research, this study aims to identify EFL students' perceptions and preferences of online listening materials and effective ways of learning using mobile learning applications in listening classes across gender. More specifically, this study is guided by these research questions:

1. Which kinds of mobile learning applications are used by the students across gender in listening classes?
2. Why do the students tend to use those mobile learning applications in listening classes?
3. How do the students perceive the use of mobile learning applications across gender in listening classes?

2 RESEARCH METHOD

This descriptive study involved thirty-five male and eighty-five female students of English Department of a public university in Banten Province, Indonesia. The students were selected because they used mobile learning applications purposefully in the listening courses they were taking. An online questionnaire as the main instrument for data collection was utilized. Moderation by experts was employed to ensure valid research instruments; improvements were made based on the experts' feedback. The questionnaire was adapted from some previous studies (Darsih & Asikin 2020; Klimova & Polakova 2020; Laghari et al. 2017); the questionnaire consisted of ten open-ended items covering students' demographic data, application preferences, and reasons for choosing and using the applications.

The data collection procedure started with getting the students' consent to follow the research process. The students' responses to the questionnaire were classified based on gender and the presented into diagrams referring to the research questions. The responses were carefully examined by relevance to support the descriptive findings.

3 RESULTS AND DISCUSSION

The data presented in Figure 1 reflect the overall results of the first questionnaire item. The students mostly used seven kinds of mobile learning applications; they were Youtube, BBC, Podcast, Edmodo, Duolingo English Learning Practice, and Vodcast. Related to the category for the mobile learning applications in this study, they were grouped into five categories: (1) dictionary or translation; (2) language practice; (3) video-based language learning; (4) interactive language learning; and (5) online courses.

Regarding language practice category, there were five males and twenty-three females students (23.3%) who used Duolingo as their mobile learning applications. Next, eleven males and twenty-six female students (30.8%) used Edmodo. The third category was video-based language learning. In this category, thirty-two male and seventy-six female students (90%) used YouTube. The results can be interpreted that mostly the males and the female students use their mobile learning applications as a tool and entertainment in language learning, especially in the listening class.

Figure 2 shows the results of the next questionnaire item. The first position with the statement of 'Its flexibility to be used anytime and anywhere' was chosen by 94 participants (78.3%), comprising twenty-six male and sixty-eight female students. Second, the statement 'As a useful tool of learning

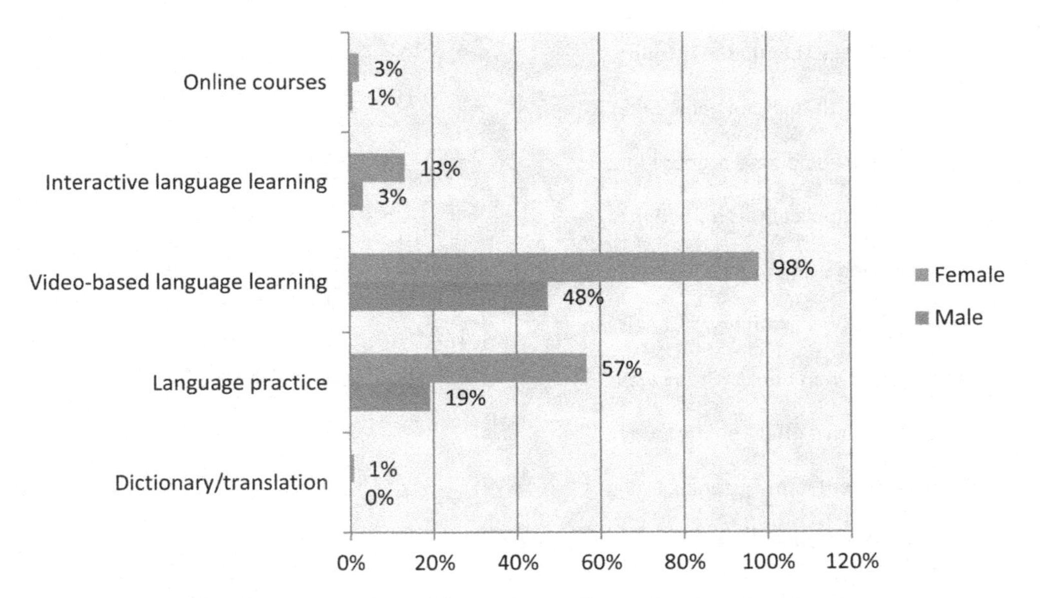

Figure 1. The students' preferences for mobile learning application based on the categories.

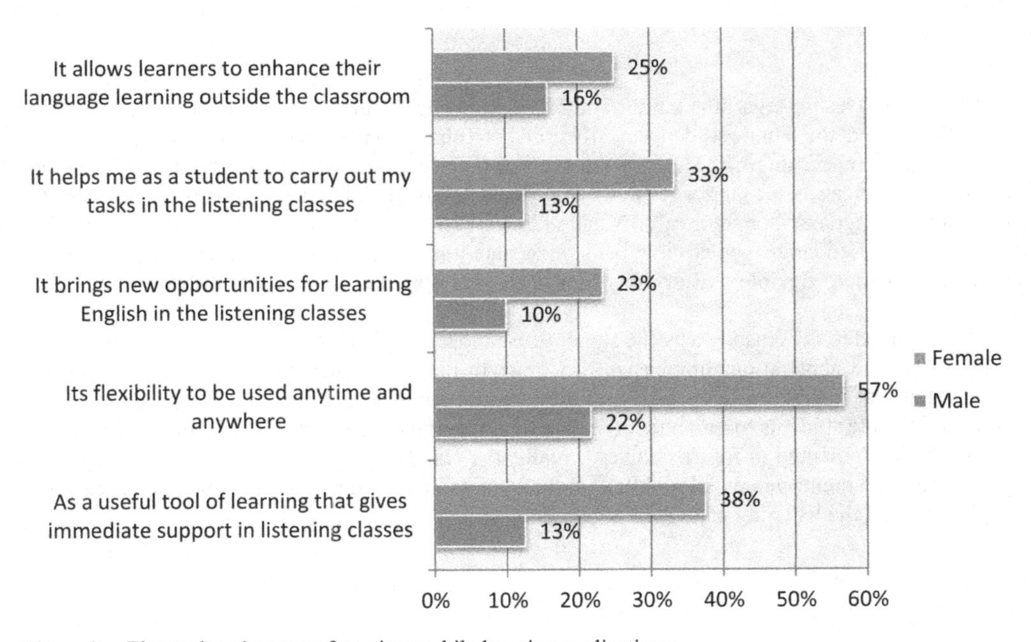

Figure 2. The students' reasons for using mobile learning applications.

that gives immediate support in listening classes' was chosen by sixty participants (50%), consisting of fifteen male and fourth-five female students.

Figure 3 demonstrates that the participants' perceptions of mobile learning applications in the listening class are related to usefulness, enjoyment, relative advantage, mobility, self-efficacy, and intention. Generally, female students perceived more usefulness of using mobile learning applications than male students. In particular, both female and male students reported that using mobile learning applications was more fun, less stressful, and more accessible than books.

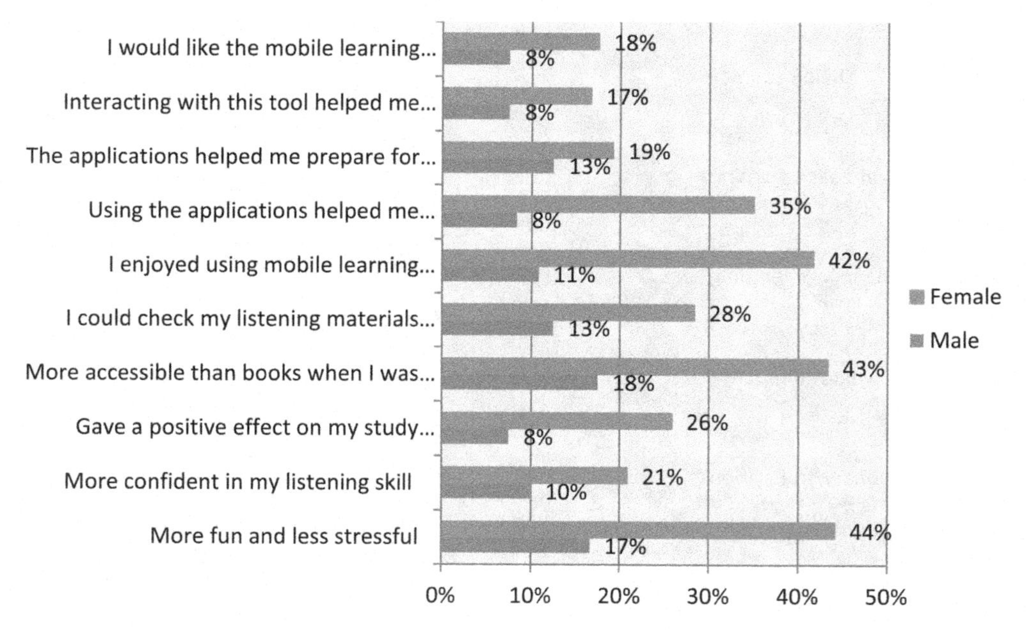

Figure 3. The students' perceptions of using mobile learning applications.

The research results regarding kinds of mobile learning applications across gender in the listening class showed that the female students mostly used YouTube and video-based language learning. The students' preferences on using Duolingo to practice their learning listening confirm that Duolingo is a useful application for languages that can provide practical and systematic steps for learners to learn a new language independently (Nushi et al. 2017). Related to YouTube and Podcast as part of the video-based language learning, the students are more likely to watch the videos on YouTube because it is more flexible and exciting, and it can be shared with their friends (Wang & Chen 2020).

Then, related to the reasons why the students use those mobile learning applications, the results show that those applications emphasize flexibility to be used anytime and anywhere. It can be interpreted that interacting with these mobile learning applications as a tool that helps both the male and female students to remember the listening materials and the mobile learning applications should be implemented in future courses. Finally, the results of this study show that the students perceived enjoyment, perceived mobility, and associated positive perceptions of mobile learning applications in the listening class across gender.

4 CONCLUSION

This study demonstrates that the male students use more than one mobile learning application than the female students based on their preferences in video-based language learning as a tool for accessing online listening materials and as an entertainment tool to support them in the listening class. Then, this study shows that the flexibility and accessibility to the listening materials as well as immediate support from the applications in the listening class are two contributing factors of the students' reasons for using mobile learning applications. They also experience more fun and feel less stressed.

REFERENCES

Bembenutty, H., & White, M. C. (2013). Academic performance and satisfaction with homework completion among college students. *Learning and Individual Differences, 24*, 83–88.

Darsih, E., & Asikin, N. A. (2020). Mobile Assisted Language Learning: Efl Learners Perceptions Toward The Use Of Mobile Applications In Learning English. *English Review: Journal of English Education, 8*(2), 19.

Gafni, R., Achituv, D. B., & Rachmani, G. J. (2017). Learning Foreign Languages Using Mobile Applications. *Journal of Information Technology Education: Research, 16*, 301–317.

Gangaiamaran, R., & Pasupathi, M. (2017). Review on Use of Mobile Apps for Language Learning. *International Journal of Applied Engineering Research, 12*(21), 11242–11251.

Jia, C., & Hew, K. F. T. (2019). Supporting lower-level processes in EFL listening: The effect on learners' listening proficiency of a dictation program supported by a mobile instant messaging app. *Computer Assisted Language Learning*, 1–28.

Jumaat, N. F., Tasir, Z., Lah, N. H. C., & Ashari, Z. M. (2018). Students' Preferences of m-Learning Applications in Higher Education: A Review. *Advanced Science Letters, 24*(4), 2858–2861.

Klimova, B., & Polakova, P. (2020). Students' Perceptions of an EFL Vocabulary Learning Mobile Application. *Education Sciences, 10*(2), 37.

Laghari, Z. P., Kazi, H., & Nizamani, M. A. (2017). Mobile Learning Application Development for Improvement of English Listening Comprehension. *International Journal of Advanced Computer Science and Applications, 8*(8).

Nushi, M., Eqbali, M. H., Boulevard, D., & Tehran, E. S. (2017). *Duolingo: A Mobile Application To Assist Second Language Learning. 17*(1), 89–98.

Reychav, I., & McHaney, R. (2017). The Relationship Between Gender And Mobile Technology Use In Collaborative Learning Settings: An Empirical Investigation. *Computers & Education, 113*, 61–74.

Viberg, O., & Grönlund, Å. (2013). Cross-cultural analysis of users' attitudes toward the use of mobile devices in second and foreign language learning in higher education: A case from Sweden and China. *Computers & Education, 69*, 169–180.

Wang, H., & Chen, C. W. (2020). Learning English from YouTubers: English L2 learners' self-regulated language learning on YouTube. *Innovation in Language Learning and Teaching, 14*(4), 333–346.

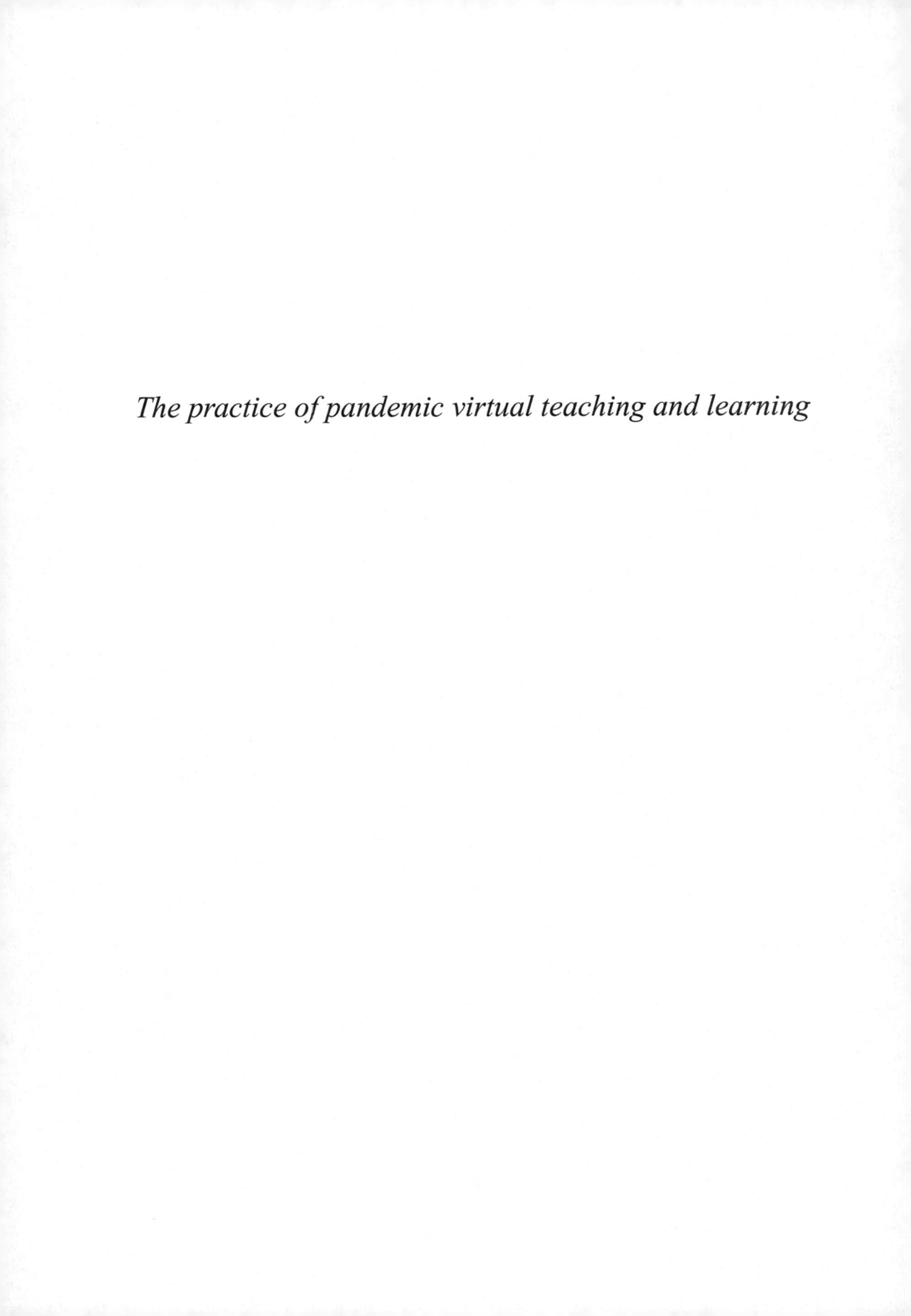

The practice of pandemic virtual teaching and learning

Post Pandemic L2 Pedagogy – Adi Putra & Arifah Drajati (Eds)
© 2021 Taylor & Francis Group, London, ISBN 978-1-032-05807-8

Online learning during Movement Control Order (MCO): Benefits and challenges

Wan Noor Farah, Wan Shamsuddin, Siti Hafizah Daud and Mashita M. Zin
INTI International University, Malaysia

ABSTRACT: This article explores the overt perceptions of 163 students who experienced the transition of online learning from face-to-face classes during the Movement Control Order (MCO) beginning from 18 March 2020 until 24 April 2020. An online survey was constructed based on Kearns (2012) with some alterations to suit the Malaysian pandemic context. The findings suggest that over 70% of the respondents believed that online class is a good alternative and it has helped them to learn during the quarantine period. They described online learning as flexible, convenient and autonomous. Despite that, the students also pointed out several drawbacks of online classes mainly on their poor internet connection and lecturers' lack of online teaching skills. All in all, this study concludes that most students had positive perceptions of online learning. This study also allows the educators to reflect on their teaching during this pandemic and continue to improve on their online teaching.

Keywords: online learning, pandemic learning, emergency remote teaching, COVID-19 pandemic

1 INTRODUCTION

Just like any other fields, COVID-19 has impacted education in many different ways. According to a report by the United Nation (2020), the pandemic has impacted 94% of the world education that led to the closure of many learning institutions. The dire situation in Malaysia in mid-March 2020 has caused the authorities to impose the Movement Control Order (MCO). Though the pandemic has caused people to stay at their own respective houses, this does not stop learning from occurring at the comfort of one's abode. Many tertiary institutions have opted for distance learning as an alternative to ensure that learning still takes place regardless of the quarantine. This has really put online learning to the test as it is the most viable option during the quarantine. This is because the debate on whether online learning can really provide almost the same quality of education with face-to-face learning has always been a long debate amongst the educators. Before moving into that debate, the definition of online learning will firstly be explained in the next section, followed by transition of teaching method which has been practiced by tertiary educational institutions. Thus, it will give a clear perspective of the online learning in this study.

2 LITERATURE REVIEW

2.1 *What is online learning?*

Thanks to the proliferation of new information and communication technologies (ICT), many tertiary institutions nowadays are equipped with platforms for online learning. According to Guri-Rosenblit (2005), the history of online learning can be traced back to a few centuries ago in response to the rapid development of ICT and demands from the new generation. Therefore, it is interesting

DOI 10.1201/9781003199267-3

to note that online learning emerged concurrently with the development of new technologies and responding to new generation's learning styles. Online learning, according to Gotschall (2000) can be defined as any type of learning that uses computer networks and possesses four main characteristics. Online learning: 1) is delivered online, 2) face-to-face sessions are not necessary, 3) has various communication tools and 4) contains different multimedia platforms like videos and PowerPoint slides. Several terminologies that might be associated with online learning are distance learning, distance education and distributed learning which may have been used interchangeably. Another similar term, E-learning is widely used to entail similar concepts. Fee (2005) defined the term "e-learning" as "any learning that involves using the internet or intranet (p.5)."

2.2 *Transition from face-to-face to online classes*

Lee and Chan (2007) point out that the quality of education between face-to-face learning can never be replaced with online learning because of the lack of physical presence which later impedes the communication process. Some researchers believe that face-to-face learning provides the solitary and emotional connections which are not evident in online learning (Kreijsa et al. 2003). Besides, many expressed their concerns on the ability of the students to engage in online learning as it is completely opposite to the conventional way of teaching especially on the lack of face-to-face contact (Simonson et al. 2009; Wagner & Hugan 2011).

In contrast, more recent research is prone to the argument that online learning has evolved and can provide a better learning environment as it promotes self-directed learners (Geng et al. 2019). This argument was made based on the hallmark of effective education that provides means for students to participate in activities such as open dialogue, debate, negotiation and agreement which apparently what online learning managed to offer through online discussion and other platforms such as social media. Zhan et al. (2011) state that online learning offers a more relaxed environment for the students as compared to face-to-face and far more convenient for the learning process. Thus, the student's distance learning environment is experiencing vast changes and it is pivotal for lecturers to comprehend the drives of students' learning.

The pandemic provides an opportunity for the tertiary institution to embark in fully online classes by utilising the use of a Learning Management System (LMS) such as Blackboard. Blackboard is one of the prominent LMS providers and has received positive feedback worldwide where Blackboard served 16,000 clients over 90 countries with almost 100 million users (Blackboard 2017). It is used as an interface or archive of information that provides lecture materials and resources. Blackboard also served as a tool for two-way communication between the students and the faculty members using different platforms such as announcements, discussion boards, mails and podcasts (Alokluk 2018).

Though much research has been done to focus on online learning in general, there is a general paucity of academic writings that focus on the usage of any learning platform during the quarantine. Since COVID-19 is a relatively recent occurrence, this study will attempt to probe students' perceptions on the usage of Blackboard during this pandemic. Therefore, the following research questions are used for current study:

1. Is the Learning Management System Blackboard a good alternative to replace face-to-face learning during this MCO?
2. What are the benefits of online learning during the quarantine?
3. What are the challenges faced by the students when doing online learning during the quarantine?

3 RESEARCH METHOD

This study employs both quantitative and qualitative approaches in which the first part of the online survey employs a five-point Likert Scale and the second part of the survey employs a more qualitative approach with the open-ended questions. To demonstrate the internal reliability of the

Table 1. Frequency for students' perception for Blackboard as an alternative during MCO (N = 163).

Question	Yes	No	I don't know
Do you agree that lecturers should use Blackboard as an alternative to face-to-face learning during MCO?	123	18	22

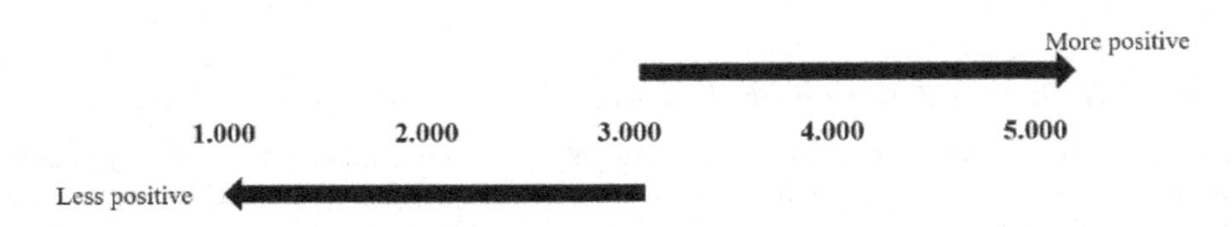

Figure 1. *Cut-off point for the Likert Scale used.*

Table 2. Students' perception on learning through Blackboard (N = 163).

Statement	Mean
Do you think that using Blackboard during this MCO is helping you as a student to learn?	3.472

*Note: 1 is used to indicate strongly disagree and 5 is used as strongly agree

questionnaire used, Cronbach Alpha was performed using SPSS Software. The internal consistency coefficient of the questionnaire was recorded as .847. According to Fraenkel et al. (2012), a reliable instrument should have the Cronbach Alpha reading above .700. Since the Cronbach Alpha reading for present study is .847, this suggests that the questionnaire is reliable. This study involves 163 students at a private university who are experiencing the transition from face-to-face learning to online learning during the lockdown because of COVID-19. 128 of the students are diploma students and 35 of them are bachelor's degree students. Majority of the participants are local Malaysian students and 32 of them are international.

4 FINDINGS AND DISCUSSION

4.1 *Research question 1: Is the learning management system, blackboard a good alternative to replace face-to-face learning during this Movement Control Order (MCO)?*

To answer the first research, the students are required to respond to the question "Do you agree that lecturers should use Blackboard as an alternative to face-to-face learning during MCO?" with a five-point Likert scale (1 as the strongly disagree and 5 as strongly agree). Table 1 illustrates the overall mean for the question.

From Table 1, it is apparent that most of the respondents believed that the learning platform Blackboard should be used as an alternative during the quarantine period. This suggests that most of the students believe that learning should still take place during the quarantine and Blackboard is the best platform to do so. This finding is particularly interesting as the decision for continuing the lectures during the pandemic received dichotomous views. This study found that most of the students agree that Blackboard is the best platform to study during the quarantine.

To further probe the students on the issue, they were asked if Blackboard helped them to learn during this pandemic. The students were asked to rate on a scale of 1 to 5 with 1 as least positive and 5 as most positive. To analyse this, a cut-off point of 3.000 will be used. In other words, mean that is below 3.000 will be considered as least positive, whereas mean that is above 3.000 is considered as more positive perception.

From Table 2, it is suggested that the overall mean of the students' responses on whether Blackboard has helped them to learn during the quarantine, the mean is 3.472. This is considered as a positive perception and attitude towards the use of Blackboard. A study done by Al-rahmi et al. (2015) showed that most university students have positive attitudes towards online learning and they are willing to learn using online platforms. This concurrent with the findings of current research indicates that most students believe that online learning is helping them to study during this quarantine period.

4.2 *Research Question 2: What are the benefits of online learning during the quarantine?*

To answer the second research question, the findings from the second part of the survey is utilised. Thematic analysis is used to see emerging themes for the question. From the survey it is found that students think that online learning brings benefits in terms of: 1) convenience and 2) user-friendliness.

1. Convenience
From the analysis, it is found that most students think that online learning is very convenient. This can be illustrated from the responses in the survey questions as seen below:

"convenient for study" – Student 63
"Able to catch up with syllabus" – Student 66
"It helps a lot as we can't travel all the way to Uni..." – Student 6

"Everything is uploaded on it and it is so convenient in finding what we need" – Student 102

One of the perks of online learning is the convenience, according to most of the responses. This is because students are able to access the course materials and perform learning tasks at their own convenient time and place.

2. User-friendly
Another occurring response from the students is the user-friendly interface that makes Blackboard one of the advantages of online learning. This can be seen from the responses as below.

"Easy to contact with lecturer" – Student 36
"Easy to use as I have been using it ever since the start of my studies" – Student 2

Most of the students are quite familiar with Blackboard as Blended Learning is practiced at the institution. This provides the students with some exposure to some of the tools embedded in Blackboard. As mentioned by Student 2, students have been using Blackboard since the beginning of their study, so using Blackboard during the quarantine is quite effortless for them.

4.3 *Research Question 3: What are the challenges faced by the students when doing online learning during the quarantine?*

Similar analysis was done to answer the third research question. The three main challenges faced by the students when doing the online classes are: 1) Internet connection 2) Difficulty in concentrating and motivation 3) Instructor's online teaching skills

1. Internet connection
To no surprise, internet connection is the most common response that was mentioned by the respondents.

"Sometimes network problems raise concerns" – Student 80
"Connection problems" – Student 83

Connection problems are not foreign in online learning discussion. Islam, Beer and Slack (2015) refer this as technological challenges that refer to the external factors that are not related to academic matters but highly impactful in online learning. For many developing countries, this is one of the biggest issues according to Ngampornchai and Adams (2016).

2. Difficulty in concentrating and motivation
Another highly mentioned issue when learning online is related to students' drive to online learning. Following are some of the responses from the students.

"Not paying attention" – Student 89
"Internet not strong sometimes and no motivation" – Student 92
"Morning classes" – Student 28
"Motivation to pay attention to teacher" – Student 97
"We have to practise discipline for ourselves" – Student 12

Motivation and concentration are the few challenges of online learning that are rooted from the learners' perspective. Much research has emphasized this such as Keengwe and Kidd (2010) that mention learners' attitudes and motivation in learning online. Nelson and Thompson (2005) highlight that procrastination is also one of the challenges in online learning.

3. Instructor's online teaching skills
One interesting finding from the survey which is instructor's skills to teach online is another challenge for online classrooms. Recorded responses for this include:

"Inability to have teacher explain certain things one-to-one" – Student 22
"Don't understand my teacher" – Student 34
"Cannot understand what the lecturer explained in the recorded slide" – Student 48

It is undeniable that teaching online is not an easy task. However, according to the survey, students found it difficult for them to understand the lesson via online learning. This can stem from the different teaching styles exhibited by the lecturers. Keengwe and Kidd (2010) note that lack of technical and IT knowledge from the lecturers may hinder the process of effective online teaching. Not only that, Keengwe and Kidd also report that several educators are quite reluctant to attend related training and support mainly due to the overwhelmed workload and class preparation.

5 CONCLUSION

In short, the current study managed to inquire the students' perspective on the use of Blackboard as LMS provider during the quarantine. Generally, the majority of the students believe that Blackboard should be used to replace the face-to-face lessons during the quarantine. There is not much difference on the opinion of Blackboard usage between local and international students on this matter as well. The students also believed that Blackboard has helped them to learn during the pandemic and highlight several strong points of Blackboard which include convenience and user-friendly. However, this study also found several drawbacks of online learning during the quarantine which are connectivity issues, motivation to concentrate in online class and the lecturers' competence in conducting online lessons. Therefore, it is important for the instructors to consider these factors when conducting online classes. This paper suggests that instructors are encouraged to keep improving their teaching style for online classes by exploring more interactive online activities and focusing on student engagement and motivation. These areas could be explored more in the future to improve the efficiency and the effectiveness of online learning. Hence, both instructors and students would gain benefits from the future research and be able to cope with the rapid technological change.

REFERENCES

Al-rahmi, W. M., Othman, M. Sh. & Yusuf, L. M. (2015). The effectiveness of using e-learning in Malaysian Higher Education: a case study Universiti Teknologi Malaysia. *Mediterranean Journal of Social Sciences.* 6(5), 625–637.

Alokluk, A. (2018) The effectiveness of the Blackboard System, uses and limitations in information technology. *Intelligent Information Management*, 10(6) 133–149.

Fee, K. H. (2005). Delivering e-Learning: a complete strategy for design application and assessment. London and Philadelphia: Kogan Page.

Fraenkel, J. R., Wallen, N. E. & Hyun, H. H. (2012). How to design and evaluate research in education. New York: McGraw-Hill.

Gotschall, M. (2000) E-learning strategies for executive education and corporate training. Fortune, 141, S5–S59.

Guri-Rosenblit, S. (2005). Distance education and e-learning: not the same. *Springer*, 49(1), 467–493.

Islam, N., Beer, M. & Slack, F. (2015). E-learning challenges faced by academics in higher education: a literature review. *Journal of Education and Training Studies*. 3(5), 102–112.

Keengwe, J. & Kidd, T. T. (2010). Towards best practices on online learning and teaching in higher education. *MERLOT Journal of Online Learning and Teaching*, 6(2), 533–541.

Kreijnsa, K., Kirschner, P. A., & Jocherms, W. (2003). Identifying the pitfalls for social interaction in a computer-supported collaborative learning environment: a review of the research. *Computers in Human Behaviour*, 19(1), 335–353.

Lee, M. J. W. & Chan, A. (2007). Reducing the effects of isolation and promoting inclusivity for distance learners through podcasting. T*urkish Online Journal of Distance Education*, 8(1), 85–105.

Ngampornchai, A. & Adams, J. (2016). Students' acceptance and readiness for e-learning in Northeastern Thailand. *International Journal of Educational Technology in Higher Education.* 13(1), 1–13.

Simonson, M., Smaldino, S., Albright, M., & Zvacek, S. (2009). Teaching and learning at a distance: foundations of distance education. Boston: Pearson Education.

United Nation. (2020). Policy brief: education during covid-19 and beyond. UN: Not given.

Wagner, R., & Hugan, J. (2011). Relative performance of English Second Language students in accounting courses. *American Journal of Business Education*, 4(5), 31–38.

Zhan, Z. H., Xu, F. Y. & Ye, H. W. (2011). Effects of an online learning community on active and reflective learners' performance and attitudes in a face-to-face undergraduate course. *Computers & Education*, 56(4), 961–968.

Post Pandemic L2 Pedagogy – Adi Putra & Arifah Drajati (Eds)
© 2021 Taylor & Francis Group, London, ISBN 978-1-032-05807-8

Zoom's screen sharing and breakout rooms in teaching reading online

Tien Thinh Vu & Diem Bich Huyen Bui
International University – VNU HCM, HCMC, Vietnam

ABSTRACT: This research study investigated the students' attitude and learning efficiency in learning IELTS reading skills via Zoom's online learning platform. A total of fifty students at the pre-intermediate level were selected, then divided into two groups for an eight-week training period. An experimental design was employed, utilising the screen sharing and breakout room functions on Zoom application. The only difference was the high rate of students' interaction and the strong focus on student-centeredness in all class meetings for the experimental students. Findings from test scores and the questionnaires given at the end of eight weeks reported effectiveness and positive attitudes towards the interactive teaching style. The outcomes of this study shed light on the belief that focusing on the learners is an effective approach to online teaching in the pandemic period.

Keywords: Online learning, student-centered, teaching reading, technology, Zoom

1 INTRODUCTION

The first patient of the Coronavirus was found in Vietnam at the end of January 2020 (from the Ministry of Health). Foreseeing the serious situation with unpredictable risks, the government commanded all the public schools and training institutions to cease the back-to-school after the Lunar new year and converted to online learning. This decision caused great concerns among educators, parents, learners, and all other stakeholders, especially those at higher education levels, where students paid a high tuition fee and asked for high standards of training facility and quality. Although a quick step on the use of online learning platforms such as MS Teams, Google meet, Zoom, Hangouts was implemented, there remained controversies over which methodology to apply and how to maintain teacher-student and student-student interaction to get learners the most benefits via the online learning mode. This study explored the student-centered approach and the functions to boost interaction via the Zoom learning platform and examined learners' reactions towards this learning style.

2 LITERATURE REVIEW

Throughout the course of teaching and learning, two main approaches, which are teacher-centeredness and student-centeredness, have been discussed and analysed by educators and researchers.

Traditional teacher-centered pedagogy is defined as the practice when the teacher is mainly responsible for transferring knowledge to students (Lancaster 2017). The teacher acts as the dominant leader to establish and enforce rules in the classroom. He makes plans of all the activities, states and explains the lesson objectives, asks questions and provides responses or feedback to students, and summarises the lesson content at the end of the lesson. Student-centered approach is

DOI 10.1201/9781003199267-4

a method of teaching in which the student is in the center of the learning process (Lynch 2010), or in other words, learners are actively involved in their own learning processes. Developing on the perspective that the student should gradually take control of the learning process, major characteristics of this approach are listed, including helping learners become aware of the learning processes and strategies, providing learners with opportunities to master the second language, boosting collaboration and interaction between learners and creating contexts for learners to investigate the language (Lancaster 2017).

In recent years, there has been a paradigm shift to learner-centeredness. It is found that interaction plays an important part in the success of teaching and learning, especially in the time of pandemics. These two notions have become the targets of many scientific research studies to suggest a path to better quality teaching and learning when holding classes via online platforms is unavoidable. A meta-analysis of the questionnaire with 799 university students examining three key types of interaction in online learning: student-content, student-teacher, and student-student revealed that these forms of interaction were "significantly related to students' self-efficacy for learning and course satisfaction" (Cho & Cho 2017, p.79). Lak et al. (2017) conducted a study with 120 10-to-16-year-old EFL learners in an experimental model with teacher-centered and learner-centered approaches and concluded that "learner-centered instruction was more effective than teacher-centered instruction in improving learners' reading" comprehension performance (p.8). In 2018, Martin and Bolliger surveyed 155 online students at eight universities around the USA. They found that working collaboratively with peers via online communication tools was selected as the most beneficial engagement strategy. Lai et al. (2019) investigated 62 university students in an 8-week pre-test post-test design. The experimental participants were engaged in an online learning community where they could interact with their peers. The finding showed that "higher interaction learners from the online learning community revealed better learning achievement and student engagement" (p.66). A survey at Midwestern university to evaluate the influence of presence and interaction on online learning reported that teaching presence and learner-instructor interaction were the most influential factors to effectiveness in teaching and learning (Kyei-Blankson et al. 2019).

3 RATIONALE AND RESEARCH GAP

To respond to the call from the government to switch to online learning, International University, a public university in Vietnam, made a quick decision and started to use the paid version of Zoom application in an effort to utilise all the functions provided by the application's developer. However, many lecturers, especially those who belonged to the older generation, were still taking the first steps in online teaching and decided to be "on the safe side." They chose to apply the traditional teacher-centered approach with more control of the class, hoping that the teaching session would go on as planned, rather than taking risks using the learner-centered approach, attempting to use the functions that would confuse and lead them to unexpected circumstances. In the meantime, students were required to pay the tuition fee, with no difference to the normal face-to-face (F2F) learning. They then requested an online teaching quality similar to that offered by the traditional F2F learning. In fact, the situation was a reflection of what was happening at the beginning of the Covid-19 pandemics in the field of education in Vietnam.

Until now, in Vietnam, almost no research findings, particularly teaching reading skills, have been officially published, exploiting the benefits of using Zoom in teaching university students with the application of learner-centered approach and an emphasis on interaction. This research study was an attempt to determine whether the use of a student-centered approach with a strong emphasis on the active role of the learners could help students achieve high scores in the IELTS reading comprehension test and find out students' attitude towards this teaching and learning style.

The objectives of this research study were to investigate (i) the relationship between the student-centered approach with the intensity of learners' interaction (via Zoom breakout rooms and screen sharing) and students' performance the IELTS Reading test and (ii) the attitudes of students towards this teaching and learning style. There are two research questions, as follows:

(i.) Do the experimental group students achieve higher IELTS Reading post-test scores than those in the control group?
(ii.) What are the students' perceptions of this teaching and learning mode?

4 RESEARCH METHOD

4.1 *The sampling*

With permission from the school, the population of the study was students from the Intensive English program, level 2 (pre-intermediate). These students were freshmen or had studied English for at least one semester. They had achieved IELTS 5.0 and wished to get 5.5 to pass to level 3 or 6.0 to be eligible to start to study the courses in their major.

Table 1. Descriptive statistics of the pre-test.

Variable	N	Mean	StDev.	Min.	Max.	Range
Pre-C	25	41.92	2.75	38	45	7
Pre-E	25	41.64	2.74	38	45	7

* *Pre-C: pre-test of the Control group* * *Pre-E: pre-test of the Experimental group*

From two classes, one with 28 students and the other with 27 students, students were asked to do an IELTS reading pre-test. Upon collecting results, the raw scores (0–40) were converted to an 0–9 IELTS band score. The conversion to IELTS band score was a necessary step to confirm participants' reading proficiency before the start of the treatment. As International University ran a 100-point scale policy, the fundamental rule of proportions was applied to convert the pre-test score one more time to the school's 100-point scale.

Based on the scores of the pre-test, 25 students of one class were selected and named as the control group (N = 25, M = 41.92, SD = 2.75), and 25 students of the other class named as the experimental group (N = 25, M = 41.64, SD = 2.74).

4.2 *Procedures*

The control group was mainly dominated by the teacher-centered approach with some modifications to modern language teaching. The teacher played the most important role in the learning process, including the curriculum and the pace of learning (Dole et al. 2016). There were phases in the lessons when students raised their voices to ask and answer questions. Actually, students could only speak up with the teacher's allowance, which helped put the class under control. Based on the finding that interaction plays an important role in achieving students' satisfaction and improving progress (Lin et al. 2017), the learner-centered pedagogy was applied for the Experimental group with more agency and involvement on the part of learners. Harmer (2011) believed that students should have control over what happened in the classroom. If students were given the power and right to act and control the activities, they would be more motivated in the whole course of learning. The two functions mainly used for the experimental students were the screen sharing and breakout rooms.

The first class meeting was reserved for asking for students' agreement to join the study. They did a 60-minute IELTS reading pre-test, based on which participants were selected. The selected names were not announced to the class to avoid any unexpected behaviors from any class members throughout the study. From the second class meeting until the end of the study, the class procedures designed for each group ran quite as a routine. All the class meetings followed the Pre, While,

Post model (Nunan 1991). The main difference in the lesson flow of the groups lay in the active role of the students with more frequent use of Zoom breakout rooms and screen sharing functions in the experimental group. In the other group, the teacher-centered pedagogy handed the teacher the highest authority, which helped take good control of the class (Lak &Parvaneh 2017). The control group employed the teacher-centered approach with the teacher sharing the screen, giving definitions for new words, presenting new skills, asking students to read individually, eliciting answers, and summarising key points at the end of the lesson. Nevertheless, the experimental group, with an emphasis on the learner-centeredness, aimed to make students active members, taking control of their studies. Students had discussions and sharing about new words, reading techniques, and comprehension questions in breakout rooms. They shared screen to give answers or presentations. The teacher just facilitated the learning process by giving instructions, introducing new vocabulary, confirming reading strategies, and answers for reading tasks. At the end of week 8, students of both groups did the IELTS reading post-test, and those in the experimental one to complete the questionnaire to find out the effectiveness and the students' perception after the training period.

4.3 *Research instruments*

Instruments included (i) the Reading pre-test and post-test, each with three passages and 40 questions in the IELTS standard format, and (ii) the questionnaire with seven questions aiming to collect students' attitude towards the student-centered approach and the interactive learning style via Zoom's screen sharing and breakout room functions, and the reasons for holding certain attitudes.

4.4 *Data collection procedure and data analysis*

Data for analysis were collected in two stages. The first was in the first class meet (pre-test) and the others were in the last week of the experiment (post-test and questionnaire). A shortened link was provided to give students quick access to the questionnaire, which could be completed on desktop, laptop, Ipad, or smartphones. Descriptive statistics of the scores of the pre-test and the post-test were created with the statistical tool Minitab19. To process data taken from the questionnaire, a spreadsheet was utilised due to its popularity, user-friendliness, and full capacity to run necessary statistical formulae.

5 FINDINGS AND DISCUSSION

5.1 *Post-test scores*

Statistical data of the post-test (Table 2) showed the mean scores of the control group and experimental group with 49.60 and 57.56 respectively. Individual scores of the experimental group illustrated a range of 20. The range of control group post-test scores stayed at 15.

Table 2. Descriptive statistics of the post-test.

Variable	N	Mean	StDev.	Min.	Max.	Range
Post-C	25	49.60	3.91	43	58	15
Post-E	25	57.56	6.19	48	68	20

* *Post-C: post-test of the control group* * *Post-E: post-test of the experimental group*

The visual illustrations of individual scores showed that the score lines of the post-test lay above the score lines of the pre-test, meaning all the students showed improvements.

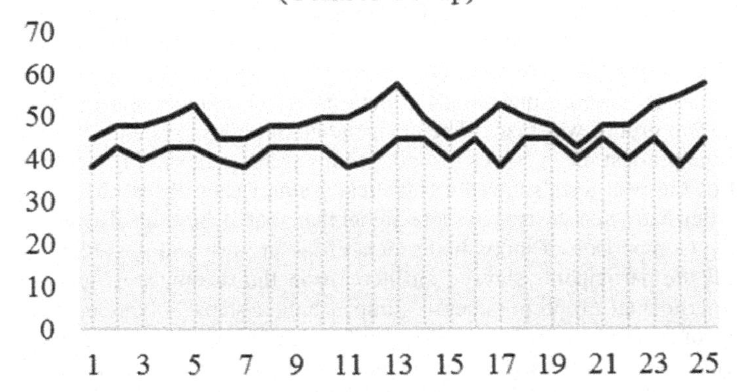

Figure 1. Pre-test Post-test Individual score comparison (control Group).
* *upper line: post-test*
* *lower line: pre-test*

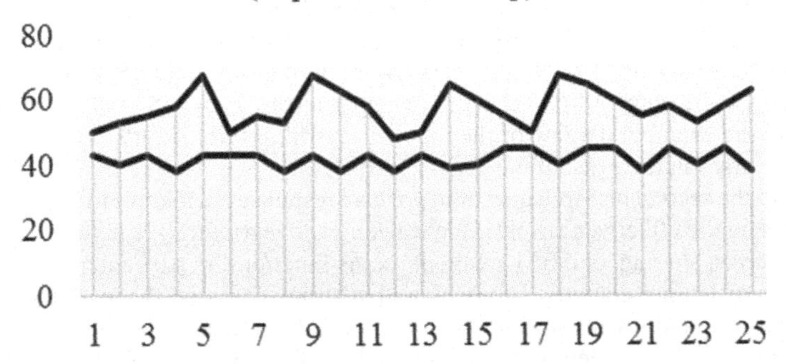

Figure 2. Pre-test Post-test Individual score comparison (experimental Group).
* *upper line: post-test*
* *lower line: pre-test*

5.2 *Questionnaire responses*

As shown in Table 3, slightly more than half of the experimental participants (52%) perceived the learner-centered and highly interactive learning style as a norm. However, a similar

Table 3. General perception of the learning style from the experimental group.

Rating □/Question □	1	2	5
Very ineffective	0%	0%	0%
Ineffective	4%	4%	0%
Normal	52%	16%	12%
Effective	20%	48%	52%
Very effective	24%	32%	36%

* Question 1: Effectiveness of student-centeredness and interactive learning style via Zoom
* Question 2: Effectiveness of students' involvement in Zoom screen sharing function
* Question 5: Effectiveness of the operation of the Zoom breakout rooms function

percentage (48%) rated this learning style as effective or very effective. Ratings on the effectiveness of Zoom screen sharing and breakout rooms functions, in particular, stayed at 80% and 88% positive respectively. However, there still existed one student who showed an attitude to questions 1 and 2 as ineffective.

To answer the first research question, an analysis of the comparison of the IELTS reading post-tests of the two groups was made. The independent Samples t-test calculated p-value >0.05, indicating that the pre-test mean scores of both groups were quite similar, while the independent Samples t-test of the post-tests showed a statistically significant difference. To be specific, the experimental group had a post-test mean score higher than that of the control group, 57.56 and 49.60 correspondingly. Comparison of individual scores of the pre-test and post-test of the two groups revealed that all the participants showed an increase in the reading post-test scores. However, those in the experimental group in general "jumped bigger steps" when compared the post-test to the pre-test scores. The improvement average of the experimental students was 15.92 (M[pre-E] = 41.64, M[post-E] = 57.56), while those in the control group obtained an improvement mean of 7.68 (M[pre-C] = 41.92, M[post-C] = 49.60). Therefore, it can be concluded that after the training period, the students in the experimental group outperformed those in the control group in the IELTS reading post-test.

Examining the scores in more detail, it is seen that the post-test scores of the control group covered a range of 43–58, equivalent to 17–23 raw scores or 5.0 to the low edge of 6.0 IELTS. This means that those at 5.0 IELTS band score failed the exam. Meanwhile, from the similar score range at the start of the study, with IELTS reading pre-test score range of 38-45, equivalent to 15–18 raw scores or 5.0 on the IELTS band score, the post-test scores of the experimental students spread over a wider range of 48–68 (19–27 raw scores or 5.5 to the threshold of 6.5 IELTS band score). More interestingly, the range 48-68 belonged to upper levels, indicating that all the experimental students passed the exam, and a few of them could even skip level 3 and, therefore, were eligible to start major courses in their discipline.

In response to the second research question, positive responses (Effective to Very effective) to the rating questions on the effectiveness of (i) the teaching and learning style in the training in general and (ii) Zoom screen sharing, and (iii) breakout rooms functions, in particular, are indications that the combination of learner-centered approach with a strong focus on the learners' interaction via Zoom screen sharing and breakout rooms functions was beneficial to learners. Responses on the reasons for their attitude revealed three major benefits of learners' involvement in the activity of screen sharing, including *easy to understand the lesson* (92%), *increase student talking time* (80%), *improve students' responsibility in learning* (76%).

Regarding the advantages of the Zoom breakout rooms function, increasing student-student interaction occupied 92%, followed by making online classes similar to face-to-face class and making students more active with 84% and 76% respectively.

Nevertheless, the findings showed one out of 25 experimental students (4.00%) considering the treatment learning style and the Zoom screen sharing activity ineffectiveness. This might be due to the fact, which was also the major drawback eliciting from questionnaire questions 4 and 7 that poor internet connection was a real barrier and annoyance to online learning. This is in line with studies that the Internet is a significant predictor of learning success and satisfaction (Chen &Yao 2016). The disruption caused by poor signals can lead to demotivation (Wu et al. 2020), boredom, and even termination from the online learning periods.

6 CONCLUSION

The findings provided grounds to claim that the learner-centered approach with an emphasis on the utilization of the Zoom screen sharing and breakout rooms functions on the teaching and learning of IELTS reading comprehension skill helped learners gain more improvement in IELTS reading, compared to the traditional teacher-centered pedagogy. However, no single method or pedagogy is perfect. Instead, a combination of methods and techniques should be used to meet the learners' needs (Harmer 2011). It is important to keep the class under control to facilitate the learning process, and to make the teacher a real facilitator who should "provide means and triggers in order to promote social presence and learner-learner interaction" (Saadatmand et al. 2017, p.72). Although the study shed light on the positive effect of the combination of learner-centered pedagogy and the intensity of learners' interaction via Zoom application, the participants were just a small group of 25 samples. A larger group consisting of hundreds of learners would be a more reliable representation on confirmation of the results. In the time of pandemics, using Zoom applications is a rising trend. However, how to utilise the functions of Zoom to make the classroom interesting and effective like the F2F mode should be the question waiting for educators and all the stakeholders involved to answer.

REFERENCES

Chen, W. S., & Yao, A. Y. T. (2016). An empirical evaluation of critical factors influencing learner satisfaction in blended learning: A pilot study. *Universal Journal of Educational Research, 4*(7), 1667–1671.

Cho, M. H., & Cho, Y. (2017). Self-regulation in three types of online interaction: a scale development. *Distance Education, 38*(1).

Dole, S., Bloom, L., & Kowalske, K. (2016). Transforming pedagogy: Changing perspectives from teacher-centered to learner-centered. *Interdisciplinary Journal of Problem-Based Learning, 10*(1), 31–45.

Harmer, J. (2011). *The Practice of English Language Teaching*. Longman: New York.

Kyei-Blankson, L., Ntuli, E., & Donnelly, H. (2019). Establishing the importance of interaction and presence to student learning in online environments. *Journal of Interactive Learning Research, 30*(4), 539–560.

Lai, C. H., Lin, H. W., Lin, R. M., & Tho, P. D. (2019). Effect of peer interaction among online learning community on learning engagement and achievement. *International Journal of Distance Education Technologies (IJDET), 17*(1), 66–77.

Lak, M., Soleimani, H., & Parvaneh, F. (2017). The effect of teacher-centeredness method vs. learner-centeredness method on reading comprehension among Iranian EFL learners. *Journal of Advances in English Language Teaching, 5*(1), 1–10. Retrieved from.

Lancaster, R. W. (2017). *A comparison of student-centered and teacher-centered learning approaches in one alternative learning classroom environment.* Parkway: Arkansas State University.

Lin, C. H., Zheng, B., & Zhang, Y. (2017). Interactions and learning outcomes in online language courses. *British Journal of Educational Technology, 48*(3), 730–748.

Lynch, D. N. (2010). Student-centered learning: The approach that better benefits students. West Virginia: *Virginia Wesleyan College.*

Martin, F., & Bolliger, D. U. (2018). Engagement matters: Student perceptions on the importance of engagement strategies in the online learning environment. *Online Learning, 22*(1), 205–222.

Nunan, D. (1991). Language teaching methodology: A textbook for teachers. New Jersey: Prentice Hall.

Saadatmand, M., Uhlin, L., Hedberg, M., Åbjörnsson, L., & Kvarnström, M. (2017). Examining learners' interaction in an open online course through the community of inquiry framework. *European Journal of Open, Distance and E-learning, 20*(1), 61–79.

Wu, W. C. V., Yang, J. C., Scott Chen Hsieh, J., & Yamamoto, T. (2020). Free from demotivation in EFL writing: The use of online flipped writing instruction. *Computer Assisted Language Learning, 33*(4), 353–387.

Post Pandemic L2 Pedagogy – Adi Putra & Arifah Drajati (Eds)
© 2021 Taylor & Francis Group, London, ISBN 978-1-032-05807-8

Peer text interaction in online classes during COVID-19 pandemic

Magfirah Thayyib
Institut Agama Islam Negeri Palopo, Indonesia

ABSTRACT: This research aims: 1) to elaborate the patterns of peer text interaction in a "WhatsApp group" class; then 2) to explore the modification/utilization of the patterns in language teaching-learning. This research is a descriptive-naturalistic study. The data was triangulated from the chats of "WhatsApp group" classes and the lecturers' notes. The number of students was 155 from five classes of English Education majors. The data was analyzed through the analytical processes, namely transcribing, pre-coding and coding, ideas growing, interpreting and conclusion-drawing. The patterns of peer-text interaction were diverse based on the course of interaction and the context of the learning activity. The dominant pattern was the "responding" goal with the "accept" and "willing" text acts. The patterns fabricated several language instruction activities for speaking and writing which are adjustable in offline and online learning. However, the practicality to draw a line from the patterns of text interaction to language instruction should be examined through further study.

Keywords: affective domain, online classes, peer text interaction, pragmatic situation

1 INTRODUCTION

Online learning which is considered the best option to cope with the situation of the COVID-19 pandemic brings about new matters and challenges. Many researchers have provided the report of the implementation of online learning during the first term of the pandemic (e.g. Ali 2020; Amin & Sundari 2020). Some researchers have also drawn a future implication from the experiences of online learning during the term (e.g. Bao 2020; Kusumawati 2020) which can be utilized in the teaching-learning process in post-pandemic time or any disruptive condition. Still, there are other details/features of online learning in pandemic time that can be exploited to be potential pedagogical supports. One of them is the pattern of peer-text interaction.

The patterns of peer interaction are varied based on the learning contexts and the auxiliary variables of the research in which the patterns are examined. In onsite learning contexts, Johnson and Belle (2016) and Damasceno (2018) have studied the learners' interaction based on their perceived outcome and the obtainable advantage. Chen (2018) investigated peer interaction in language learning according to the learners' role. In a mixed onsite and online classroom, Vu and Fadde (2013) have investigated the students' interaction based on its mode. Specifically, in text-based online classes, Coyle and Reverte (2017) and Culpeper and Qian (2019) have studied the patterns based on the interactional strategies and communicative styles. Indeed, peer text interaction in online classes during pandemic has its own distinctiveness.

From those prior studies, plenty of peer interaction patterns have been evoked but the pedagogical employment of the patterns is rather under-explored. Thus, this research aims to explore: "what are the patterns of peer-text interaction in a "WhatsApp group" class during COVID-19 pandemic?" and "what are the potential activities for language instruction based on the modification/utilization of the patterns?" The patterns of peer text interaction which occur naturally in any situation can be nurtured into language teaching-learning activities because the language is there as the medium

 DOI 10.1201/9781003199267-5

of interaction and as the object in instruction. The affective domain of the peer-text interaction particularly in the pandemic situation is also enclosed in the potential language instruction activities.

2 LITERATURE REVIEW

"Peer interaction is any communicative activity carried out between learners where there is minimal or no participation from the teacher" (Philp et al. 2014, p. 3). It is commonly utilized to mediate students' cognitive development. In various forms, it is favorable for language instruction. "Peer interaction increases learners' comfort levels and creates more opportunities for them to negotiate linguistic aspects" (Qiu 2018, p. 250). Peer interaction also occurs in the online class either virtual meeting or synchronous chat. Online chats have an equal effect with face-to-face discussion or video tools toward the students' interaction and feeling in learning (Sherman et al. 2013; Sampson & Yoshida 2020).

Alghamdi (2014) modified a variable schedule to identify the purpose of the learners' verbal interaction in a foreign language class. Whilst, Crawford et al. (2018) used corpus techniques to identify the linguistic markers of peer interactions in a second language program. Lim et al. (2018) used social network analysis to research the course of students' interaction in groups at science classes. Then, Golonka et al. (2017) developed a concept map to explore the characteristics of the learners' interactions during the text chat activities in a language training. Robles et al. (2019) used cognitive, pedagogic and affective categories to investigate the online students' interactions in a law subject at higher education. Peer interaction in an online class can be approached as it is in an offline class.

This research employs pragmatic variables and pedagogical objectives to bridge peer text interaction to language instruction activities; Leech's aspects of speech situation are combined with Bloom's affective domain. "The aspects of situation in an interaction are: 1) addressers and addressees, 2) the context of an utterance, 3) the goal(s) of an utterance, 4) the form of speech act and 5) the product of verbal act" (Leech 1990, p. 13). For the case of this research, speech/utterance refers to text chat; verbal refers to written/typed. The context is the learning activities in 'WhatsApp group'. The goal and the form of text chat are in which affective domain and subdomain are integrated. The affective domains are receiving, responding, valuing, organization, and characterization (Bloom in Isaacs 1996).

3 RESEARCH METHOD

The characteristic of this research is descriptive-naturalistic. This research used a mixed method by triangulating qualitative and quantitative data. There were five classes of English education majors in an Indonesian state university taken as the locus of this research, namely three classes in the 4th semester and two classes in the 2nd semester. The total number of students was 155. The lecturers and the students had been asked for permission about the research before the midterm exam. The data were obtained from "WhatsApp group" chat and lecturers' notes by documentation. The group chats were screen-captured while the lecturers' notes were photographed with the help of the lecturers. The data were analyzed through the analytical process which consists of transcribing, pre-coding and coding, ideas growing, interpreting and conclusion drawing (Dornyei 2007).

The peer chats were transcribed to trace the addresser and addressee, the responded and response text, also the quoted and comment text. Through pre-coding and coding stages, the goal and the form of the text chat were identified and classified into affective domains and subdomains. The prototype product of the written act was entailed in the classification. The lecturers' notes about peer interaction were also sorted and classified. Growing ideas stage was done by calculating and tabulating the patterns of peer text interaction based on their classifications. Interpreting stage was accomplished by confirming the patterns of peer text interaction with the lecturers' notes. The interpretation of the patterns of text interaction was assessed and generated into several potential activities for language instruction.

4 FINDINGS AND DISCUSSION

This section firstly presents the patterns of peer text interaction in "WhatsApp group" class. The patterns are elaborated based on the course of interaction and the contexts of learning activities. The contexts of the learning activities are question-answer session, classroom discussion, and the interactive drills session. Secondly, this section presents some potential activities for language instruction explored from the patterns of peer text interaction.

4.1 *Patterns of peer text interaction*

The total number of peer text interactions identified was 320 texts. It was 46% of the overall chats documented. The rest was text interaction that occurred between lecturer and students. Peer interaction is obvious in text-based online learning. The students' psychological aspect due to the pandemic situation is broadcasted through their text chats. The students were willing to try harder in typing the text to keep in touch with their friends during the learning process. They knew that the lecturer was there to also read the text, but they became less hesitant to address their friends because the expression was typed.

Table 1. The patterns of peer text interaction based on the course of interaction.

The pattern of peer text interaction and percentage					
		addresser-addressee		responded-response	
Direct interaction	96%	peer-to-peer	50%	one way	21%
		peer-to-group	45%	two ways	79%
		peer-to- class	5%		
Indirect interaction	4%	quoting from peer	60%		
		quoting to peer	20%		
		using peer name	20%		

The percentages of direct interaction in Table 1 reveal that the students like to address his/her friend as an individual directly. If the response is indirect, the lecturer's text or the peer's name is involved. "Quoting from peer" is when a peer text is quoted to be addressed back to him and the lecturer. "Quoting to peer" is when a lecturer's text is quoted to be addressed to a peer. In "quoting from peer" the lecturer is addressee inclusive while in "quoting to peer" the lecturer is addresser inclusive. The involvement of the teacher/lecturer in peer interaction can be initiated. The teacher may make an intervention in the "WhatsApp group" to encourage or discourage learners' interaction (Robles et al. 2019).

As can be seen in Table 2., the patterns of peer text interaction during the question-answer session are relatively similar to the ones in onsite learning. Coyle and Reverte (2017) also found that the interactional strategies of foreign language learners in online chat are similar to those in face-to-face class. There was a pattern set of "valuing" and "organization" by the group members who shared the material, "responding opportunity" goal by the floor and feedback from the group through "responding" and "valuing" questions from the floor.

The chat details of those three patterns were in formal language which needed extra work to be realized in a text. Thus, such text details were no longer appearing in the later patterns as the interaction became more vibrant. Crawford et al. (2018) identified less formal language between peers in high interaction. The last two patterns are substituting each other. The pattern set of

Table 2. The patterns of peer text interaction in the context of question-answer session.

Goals of peer interaction	Forms of text act	Text chat details
• Valuing (group vs. class interest) • Organization	assume responsibility regulate	use formal language; politely direct group/peer/class; use vocative; type a list for the class; use emoticon
Responding (opportunity)	willing; obey; engage; participate	use formal language; greet; thank; deliver a question
• Responding (question) • Valuing	willing; engage; display; devote; participate; enrich continuing desire; assume responsibility; participate; initiate	use formal language; greet; tag/type the addressee's name; use vocative; quote/re-type the question; post extra explanation; use emoticon
• Responding (opinion) • Valuing • Organization	consider; enrich; extend continuing desire; examine form judgment; relate	use vocative; appreciate and quote the opinion; deliver further comment
• Receiving (answer) • Responding • Valuing	listen > read, observe accept; consider feel, examine	thank the group/member; use vocative; assess the answer; state the act; use emoticon

"responding answer/opinion" with "valuing" and "organization" goals showed that a non-group student did indirect receiving. While the pattern set of "receiving," "responding," and "valuing" showed that the student directly received the answer.

Table 3. The patterns of peer text interaction in the context of classroom discussion.

Goals of peer interaction	Forms of text act	Text chat details
Responding (question)	accept; consider; extend	quote the question and write "2"/deliver a further question
• Responding (question) • Valuing	willing; participate assume responsibility; initiate	quote the question and answer briefly; use emoticon
• Responding (question) • Organization	willing; consider; display relate; regulate	(quote the question and) type "UP"; quote the answer and type "." (dot); direct peer to "scroll"; tell the reference
• Receiving (answer) • Responding	listen > read, observe accept; consider	thank; express feeling; use vocative
• Responding (confession/feeling) • Characterization	accept; consider; extend ready; view; approach	admit, praise, express empathy, give solution; instruct action
Responding (explanation/ instruction)	accept; display; consider	use vocative; quote the explanation and type "."; quote the instruction and encourage class/friends; use emoticon

All the patterns (sets) of peer text interaction in the classroom discussion, as shown in Table 3, involve "responding" goal. There were three patterns of "responding" questions identified. The first pattern is labeled as "basic question-responding". The second pattern of "question-responding" is a set with "question-valuing". The third pattern set is "question-responding" and "organization" domain – a combination of the first and second patterns.

Then, there were "answer-responding" with its "receiving" and "confession-responding" with "characterization" domain. Their text details were "giving solution" and "instructing action" which considered as elevated level of affective domain. The last "responding" pattern was toward a friend's explanation or instruction. In this context, many distinctive styles of "WhatsApp" chat appear as text details like quoting, using emoticons and typing symbolic statements or instructions. They are helpful in making the discussion run smoothly; facilitating the students to address their friends. The nature of the online environment led the learners to use symbols and emoticons during learning (Coyle & Reverte 2017).

Table 4. The patterns of peer text interaction in the context of interactive drills.

Goals of peer interaction	Forms of text act	Text chat details
[-] Responding (friend's phrase/clause/ sentence)	consider; prefer	quote and correct the wrong word/ structure; indirectly reject the structure
• Receiving (correction)	realize	reply and revise the structure
• Responding	accept	
Responding (friend's phrase/clause/ sentence)	accept; consider; engage; extend	quote the structure and praise/tease it using other phrases/ clause/sentence (be amused because of the meaning of the structure)
Receiving (peer name)	[to] attend, realize	type a sentence with a friend's name; tag the name

It can be seen in Table 4 that the "responding" domain appeared in both positive and negative senses during interactive drills. The "responding" was negative when the forms of text acts were "considering" a friend's sentence as wrong and "preferring" other/better structures. Golonka et al. (2017) referred to this goal as either giving the correct answer or prompting to self-correct. Toward a friend's correction, there was a pattern set of "receiving" and "responding" in which the students "realize" and "accept" the correction. The "responding" was positive when the forms of text act were "accepting" and "considering" that the structure is correct then even "engaging" and "extending" it. Because the object of the interactive drills was language, the realization of the text chats is related to language activities like typing a sentence structure, praising a correct structure, or rejecting then revising a wrong structure.

From the three contexts of learning activities in "WhatsApp group" classes, "responding" was the most dominant affective domain. Vu and Fadde (2013) found that students made comments more than they asked questions in both verbal and text modes; comment commonly represented response. Culpeper and Qian (2019) also found that affective aspects have a vital role in online language learning. The result of this research is related to those previous findings that "responding" is mostly performed by the students to affectively address their friends in chat. Peer responding can also be said as the significant affective domain in "WhatsApp group" class. The statement by Robles et al. (2019) that "WhatsApp chat group" worked to establish contact with the students' peers is acceptable for the case of this research.

The general characteristics of peer interaction that are collaborative, multiparty, and symmetrical participation structure (Philp et al. 2014) are also relevant to the findings of peer text interaction in this research. Peer interaction in text-based "WhatsApp group" class was collaborative in that the students worked together to obtain learning goals including affective goals; multiparty since the students also maintained two ways chat and used the feature that allows them to quote friend's text.

4.2 *Potential activities for language instruction*

Many indirect text interactions were done by quoting from a peer. It indicates peer resources in which the students do share opinions and feeling within the same level. From this pattern, the potential activity for language instruction is to provide the students with copious language expressions used to refer back to other's comment (question or opinion) to replace typical chat symbols like "UP," "." and "2." It might be tricky to avoid such chat convenience in synchronous text-based online learning but for language classes, it must be beneficial. For onsite language learning, this proposed activity can be in the form of a speaking drill.

"Responding" domain in the form of encouragement to friends was high. The activity suggested is providing the students with ample expressions to be used in chat to support their friends. The language provided may include the expressions/terms due to pandemics (other disruptive situations). The language might be very technical and advance but would become worth comprehensible input. For onsite learning, peer text responding can be enrolled in peer correction during writing class. The students are asked to write additional sentences to encourage their friends to improve their writing or to keep up the spirit of learning.

"Responding" in the interactive drill consisted of both language exercise response and affective response. Ideally, the response to language should be greater in interactive drills. Thus, the teacher should organize that all students participate in the drill then do peer-to-peer correction (on language structure) as an integrated activity. This integrated activity is suitable for both speaking and writing drills in synchronous online or offline language learning.

The highest percentage of "responding" domain which occurred almost with all other domains reveals the emotional/psychological aspect which can be built or stimulated. Then, interesting scaffold peer text interaction can be captured and adapted to dialogue script for offline role play. Another potential activity is that the teacher/lecturer can elicit peer interaction with affective language approach in text chat learning or direct verbal learning.

The proposed language instruction activities are similar to other forms of peer support in learning. The students' interactions during the chat activities have three types of learning support namely using partners as a resource, providing language-related assistance, and providing encouragement (Golonka et al. 2017). Another important fact is that the potential language instruction activities generated from the patterns of peer text interaction are mostly in the form of writing and speaking activities. This fact can be confirmed by the statement that text chat is a hybrid of speech and writing (Philp et al. 2014). On the other hand, it can explain the efficacy of text-based synchronous online learning like through "WhatsApp" chat.

5 CONCLUSION

The explorations of peer text interaction patterns and their utilizations as language instruction activities were feasible by combining the pragmatic aspects and the affective domain in the analyses. Diverse patterns (sets) of peer text interaction were found according to the course of interaction and the context of the learning activity. The dominant pattern was "responding" goal with two main forms namely "accept" and "willing" which confirmed the existence of affective domain in the learning. The patterns of the text interaction fabricated some language instruction activities for speaking and writing which are adjustable in offline and online learning. In any context/situation of language learning including in online text-based class, the students should be facilitated to use beneficial and qualified language through the interaction. The teacher should also maintain the affective domain in every learning activity. Then, the convenient practicality to draw a line from interaction patterns to language instruction should be further examined. Noticing other language pattern of classroom interaction in renewed mode or specialized context would be a fruitful study as well.

REFERENCES

Alghamdi, R. (2014). EFL learners' verbal interaction during cooperative learning and traditional learning. *Journal of Language Teaching and Research, 5*(1), 21–27.

Ali, W. (2020). Online and remote learning in higher education institutes: a necessity in light of covid-19 pandemic. *Higher Education Studies, 10*(3), 16–25.

Amin, F. M. & Sundari, H. (2020). EFL students' preferences on digital platforms during emergency remote teaching. *Studies in English Language and Education, 7*(2), 362–378.

Bao, W. (2020). Covid-19 and online teaching in higher education: a case study of Peking University. *Hum Behav & Emerg Tech. 2020*, 1–3.

Chen, W. (2018). Patterns of pair interaction in communicative tasks: The transition process and effect on L2 teaching and learning. *ELT Journal, 72*(4), 425–434.
Coyle, Y. & Reverte, M. J. (2017). Children's interaction and lexical acquisition in text-based online chat. *Language Learning & Technology, 21*(2), 179–199.
Crawford, W. J., Mcdonough, K., & Mercer, N. B. (2018). Identifying linguistic markers of collaboration in second language peer interaction. *TESOL Quarterly, 0*(0), 1–28.
Culpeper, J. & Qian, K. (2019). Communicative styles, rapport, and student engagement: An online peer mentoring scheme. *Applied Linguistics, 0*(0), 1–32.
Damasceno, C. S. (2018). New pathways: Affective labor and distributed expertise in peer-supported learning circles. *Communication Education, 67*(3), 330–347.
Dornyei, Z. (2007). *Research Method in Applied Linguistics*. Oxford, NY: Oxford University Press.
Golonka, E. M., Tare, M., & Bonilla, C. (2017). Peer interaction in text chat: Qualitative analysis of chat transcripts. *Language Learning & Technology, 21*(2), 157–178.
Isaacs, G. (1996). *Bloom's Taxonomy of Educational Objectives*. Queensland, Australia: Teaching and Educational Development Institute, The University of Queensland.
Johnson, Z. D. & Belle, S. L. (2016). Student-to-student confirmation in the college classroom. *Communication Education, 65*(1), 44–63.
Kusumawati, A. J. (2020). Redesigning face-to-face into online learning for speaking competence during covid-19. *International Journal of Language Education, 4*(2), 276–288.
Leech, G. (1990). *Principles of Pragmatics*. New York, NY: Longman.
Lim, S., Park, K. C., Ha, M., Lee, H., & Kim, Y. (2018). Verbal interaction types in science inquiry activities by group size. *Eurasia, 15*(7), 1–14.
Philp, J., Adams, R., & Iwashita, N. (2014). Peer interaction and second language learning. New York, NY: *Routledge*.
Qiu, X. (2018). [Review of the book *Peer interaction and second language learning: pedagogical potential and research agenda,* by M. Sato & S. Ballinger (eds)]. *Applied Linguistics,* 39(2), 250–253.
Robles, H., Guerrero, J., Llinas, H., & Montero, P. (2019). Online teacher-students interactions using Whatsapp in a law course. *Journal of Information Technology Education Research, 18*, 231–252.
Sampson, R. J. & Yoshida, R. (2020). Emergence of divergent L2 feelings through the co-adapted social context of online chat. *Linguistics and Education, 60* (2020), 1–10.
Sherman, L. E., Michikyan, M., & Greenfield, P. M. (2013). The effects of text, audio, video, and in-person communication on bonding between friends. *Cyberpsychology, 7*(2), 1–13.
Vu, P. & Fadde, P. J. (2013). When to talk, when to chat: Student interactions in live virtual classrooms. *Journal of Interactive Online Learning, 12*(2), 41–52.

Post Pandemic L2 Pedagogy – Adi Putra & Arifah Drajati (Eds)
© 2021 Taylor & Francis Group, London, ISBN 978-1-032-05807-8

Teacher's reflection in online speaking class during COVID-19 pandemic

Audrey Ningtyas and Sonya Puspasari Suganda
Universitas Indonesia, Depok, West Java, Indonesia

ABSTRACT: The sudden use of an online learning management system is creating problems for teachers. This research aims to identify the problems that occur during teaching an online speaking class in the COVID-19 pandemic era through a case study of an English teacher in Jakarta. Observations of the teaching and learning process, then interviewing the teacher are conducted to elaborate on the problems for one teacher in one of the high schools in Jakarta. At the end of the research, researchers found that the teacher is exhausted due to the lack of other teachers. Besides having technical problems, she found that students in 7th grade have distinct English proficiency for students from public school and private school. Another finding concerns the lack of motivation to participate actively in an online class. This research will give insights into the importance of teaching preparation and reflect on our own teaching to improve our teacher professional development.

Keywords: Online speaking class, online teaching, teacher professional development, teacher reflection

1 INTRODUCTION

The COVID-19 pandemic has been going on since the 31st of December, 2019. The first three months of 2020, Indonesia educated its people through news and social media to prepare for the inevitable. On March 24th, 2020, the Minister of Education and Culture announced that National Examination (UN) would not be held that year due to the risk of spreading the virus and the teaching and learning process are conducted online. Since then, the teachers use online platforms such as Skype, Google Classroom, Scola and so on. The abrupt change from the traditional classroom into an online teaching and learning process is worrying because apparently, not all teachers have the knowledge of using such online platforms. This study will investigate what problems that might emerge during online speaking class.

Preparing an online class without acknowledging the LMS (Learning Management System) would hinder the teaching and learning progress. Therefore, it is important for teachers to develop their own professional skills. There are many platforms that could be used by teachers. However, teachers needed to prepare themselves to make learning designs with such tools, and because of this pandemic, teachers are forced to use online platforms with minimum knowledge and understanding. This may lead to teachers' affective aspect of teaching. Although the platforms made the teaching and learning process easier and efficient, teachers may not be ready for such responsibilities due to their lack of knowledge and understanding (Rakes & Dunn 2015).

Teachers who are passionate in their own teaching would also be concerned about how to assess students in an online class environment. Online classes proved to have relied on students' motivation to see if the learning process successful or not (Lin & Warschauer 2015). Unmotivated students will use this chance (online learning) to procrastinate and they may not complete any tasks given by the teacher or participate actively in a virtual environment, since the teacher would not come

DOI 10.1201/9781003199267-6

to their house to check-up on them. This is truly disastrous if the teacher relied on the scores from multiple-choice tests for speaking assessment. To prevent this, teachers should emphasize the importance for students to complete their tasks and abide by deadlines by reminding them and checking on their work relentlessly. Teachers could also regularly perform a formative assessment to raise students' participation and guide them to self-reflect on what they had learned and relate them to students' everyday lives (Rakes & Dunn 2015).

This study covers the problems that emerge during the teaching and learning English as a Foreign Language, especially in speaking class, where students are required to speak English at all times. As a preliminary study, it only focused on one female English teacher in Junior High School located in South Jakarta.

1.1 *Online speaking class*

Compared to the traditional classes, where students have to face the teacher directly and can be seen by their friends, their anxiety levels will rise. They will stutter and make mistakes in traditional speaking classes. These obstacles surely would affect students' performance in practicing speaking in the future. They might become unmotivated and decide not to practice anymore to avoid mistakes and humiliation. According to Peeters (2018), an online class could reduce anxiety in speaking a foreign language. It also proved that practicing communication skills with technology could encourage students to reduce students' anxiety levels (Ataiefar & Sadighi 2017). An online class could also be conducted side by side with traditional class, research conducted by Wang (2020) proved that an online class could improve students' English conversational skills. However, teachers need to note that in order to have a successful speaking class, at least students should be exposed to authentic language settings. Marull and Kumar (2020) stated that it is important to include an authentic setting because students are not only learning the language but also the culture of the target language.

1.2 *Teacher's reflection*

Teacher's reflection derived from John Dewey (1933) who strongly believes that reflection is important for both teaching and learning. It leads to the progress of improving teaching and learning. Both teacher and students could change their strategies and behavior after the reflection. This self-regulation would immensely motivate online class participation. Teachers need to improve their professionalism continuously for the sake of the quality of teaching they want to achieve (Wong 2011), especially multicultural students (Prenger et al. 2017). Nowadays, reflective teaching has been explored into more varied types of reflection. Through writing journals, Ghorbani et al. (2020) using Farrell's (2004) framework in reflective journals. Teachers could also use a video recording of their teaching and have a self-reflection. In recent research done by McCoy and Lynam (2020), it was found that video recordings contain rich evidence that could help teachers to reflect immensely. Alemi and Tajeddin (2020) stated that teachers' reflection is always closely related to good teaching. They also identified several themes of reflections and the frequencies of each theme. In this study, researchers developed several themes to be used as the focus of reflection, such as classroom management, learners' classroom participation and responsibility, development of learners' language skills, teaching methods and techniques, raising learners' motivation and self-confidence, textbook and syllabus, classroom facilities, teachers' working conditions and teachers' classroom preparations.

If a teacher stopped improving their teacher's professional development, they jeopardize their own career and the quality of education students get from them. Reflecting on teaching does not mean teachers must solve the problems by themselves. They could discuss the problems with other teachers, find solutions with help from school or find literature online. Teachers could also attend webinars and workshops.

2 METHODS

The participant of this study is a female teacher about 26 years old, and her name would be abbreviated as NN. Since she taught in this school, she developed a reflecting habit about what happened to her in her professional field and her teaching journey for the particular school. She teaches from 7th grade until 9th grade. Because of this pandemic, the teacher was forced to teach with online platforms such as *Google Meet, Zoom, WhatsApp Messenger* and so on. Researchers wanted to focus on an online speaking English class, where students assessed orally, and the teacher needed to hear clear students' voices in order to assess objectively.

This qualitative case study tried to investigate the problems that may emerge in online speaking class during the COVID-19 pandemic and what can be done to solve the problems that emerge by suggesting several applications and programs available for students and teachers. The study includes problems gathered from the participant through semi-structured interviews with the teacher and online class observations. These questions were made based on the Alemi and Tajeddin (2020) themes of reflection. The observation guideline was also based on the same author and prepared to understand the teacher's reflections better. The semi-structured interview questions consist of 9 questions, and the interview was done via *WhatsApp Messenger* call and chats. Researchers observed the online speaking class for 2 weeks, the 7th grade observed for 1 week, and the 8th grade also for 1 week to find out more of what happened. The interviews and observations are then analyzed, and the findings are divided into four parts.

Table 1. List of semi-structured interview questions.

No.	Questions
1.	How long have you been teaching English?
2.	When did you graduate from University?
3.	Since when have you taught at this school?
4.	Which grade do you teach English?
5.	Is there any speaking or conversation class? If there is a speaking or conversation class, please elaborate on the curriculum.
6.	How do you teach the speaking or conversation class during this pandemic?
7.	How was the students' engagement on the online platform?
8.	How do you assess their speaking skill on the online platform?
9.	During this time, what does the school provide you?
10.	What are the difficulties you faced when using an online platform?

Table 2. Guidelines for online class observation.

	Objectives	
No.	For Teacher	For Students
1.	Online material	Response from teaching
2.	Proficiency in technology	Learning engagement
3.	Formative assessment	Participation
4.	Instructions	Motivation
5.	Class control	Tasks completion
6.	Assignments	Understanding the topic

3 FINDINGS AND DISCUSSION

Researchers managed to gather all the data from the interview and observations. These are several problems found after reflecting and are divided into 4 parts below guided by Alemi and Tajeddin (2020) themes of teacher reflection.

3.1 *Teacher proficiency*

Based on the interview, the participant uses *Google Meet* for teaching speaking materials, then proceeds to an application named *Scola* for assessing students for the current material. Both students and the participant, the teacher herself, could access the application from their own smartphones, and students are expected to turn in their assignments via *Scola*. Participants also use the voice note feature in *WhatsApp Messenger* to assess students' oral tests.

Judging from the observations, the participant could use these platforms smoothly and without a problem. Although participants could improve their own knowledge regarding the platforms, she could use other platforms than using the voice note feature. An Android application called *Talk and Comment* could help better than *WhatsApp* voice notes, as it does not use an internet data plan, and it does not take long to load. Instead of sending voice notes, students only need to send a link to their own recording. The recording will stay on the cloud storage for 90 days; thus, the teacher could reassess the recording. Another online platform that does not require the user to download an application is *Vocaroo*. Teachers and students could access this platform on *vocaroo.com* in the browser. One-click on the record button and then click the stop button when finished. Teachers and students could either download or send the link.

Regarding the materials, the participant had no issues in delivering via online platforms. The slides she shared with students are full colors in hopes to catch students' attention. She also provides pronunciation buttons that she could click to give authentic examples to the students.

One hour before the class starts, the participant gave students a link to Google Meet. She would wait for students until all of them are present in the video conference. After that, she began taking attendance and presenting material in English and Indonesian as not all students could understand English fully, particularly in 7th grade where she would constantly be translating her English. When the material is done, she gives instructions to students for their assignment that day via *Scola* and *Whatsapp* messages. At the end of class, she would take another attendance to ensure all students are present when she teaches.

From the observation, researchers find that the participants regularly give them formative assessments after each sub-material. Students would answer her questions based on their understanding and continue to do so. Although sometimes one of the students typed unrelated topics in the chat group, the participant could not reprimand them immediately, and it kept happening.

3.2 *School facility*

In the interview, the participant said that although she conducts online teaching, she had to come to the school and use the school's Wi-Fi connection. The school is convinced that teachers should stay in school because they do not have to buy data plans, and it is easier for teachers to discuss the materials together. All English teachers are gathered in one classroom and with one terminal cable for charging laptops.

Unfortunately, when the participant teaches, the other teachers' voice can be heard too. Since the participant's voice is rather low, it is hard for students to hear clearly what she said. Sometimes the students complained they could not hear her voice. Also, the Wi-Fi connection is being used together at the same time. It makes the traffic connection full, and the participant could be thrown out of the meeting because of the lack of signal. This brings confusion to the students, and they think the class is over.

The *Scola* application also had issues of its own. Participants experienced the delay when she had to download the assignment from students. The students also experienced difficulties when uploading their assignments. Instead of using Scola, the school committee could use other platforms such as *Google Classroom.* It is more reliable and easier to use, and teaching and learning objectives will be done timely.

3.3 *Problems reflected on teaching online speaking class*

According to the participant, students come from two distinct backgrounds: private school students and public-school students. These students from private schools and public-school have very distinguishable English proficiency. Private school students could understand what the participant says easily while the public-school students could understand nothing of what she says and teaches. This is due to the Policy made by the previous Ministry of Education and Culture, where English is no longer a subject at public school. This frustrates the participant as she has to speak both English and Indonesian and then has to be able to make all students, including the public-school alumni, speak English well. To solve this problem, the participant uses meaningful tasks close to the students' daily activities and lives. Fortunately, the strategy is a success, although some of the students are still passive. By using a different approach to online practices, students might be more interested in learning (Knapp 2018). The participant also disagreed with the school's policy to make the speaking class based on the facility they give to the students rather than their ability. She had tried to consult the facility with the school boards, but she got no answer from the school boards and followed the policy.

She also had to teach them online where bad connections occurred for both her and the students. This interrupts the learning process and it cannot be avoided. The school and curriculum committee should discuss this matter thoroughly. They could divide students with placement tests. It will at least lessen the teacher workload and make it easier to adjust the level of materials. The Participant also claimed that the teachers are expected to make their own curriculum and lesson plan for speaking class. These administrations' work takes a toll on the participant; she was confused in adjusting the materials based on the students' needs and abilities.

These issues lead to other problems. Students are not eager to learn. They become demotivated by the issues and decide not to take the assignments seriously. Many of them did not turn in their assignments, even if the participant called their parents one by one. When the class was about to start, they were expected to turn on their cameras, but only a few of them obliged. The participant could not truly assess students' oral assessment as the online platforms are not working properly and delaying the other work. Instead of having an oral summative assessment, the participant only made gap filling and matching tasks and relied on the formative assessment for students' oral assessment.

3.4 *Importance of teacher professional development*

Rotermund et al. (2017) stated that teacher professional development not only benefits teachers themselves but also boosts students' learning. As stated before, the quality of teachers' knowledge and understanding will determine the quality of education they give to students. It means, with the correct approach or strategy, teachers could improve their students' competence with ease. Teachers' without proper knowledge and understanding of the best strategy results in students' failure to achieve their learning objectives. Therefore, it is highly important for teachers to improve their professional development, and it can be done in many ways (Nurkamto 2020). Collaboration between school and the teacher is suggested to improve the school's quality itself such as attending workshops, holding a forum group discussion and so on. While the school might provide teachers workshops and webinars, teachers as the source who experienced the problems in the class should reflect on their own teaching and work together with the school's committee to solve the problems. Although, in order to make the professional development effective in teachers' practice, the developments need at least 6 to 12 months to manifest. Teachers' professional development is time-consuming, but it will be beneficial in the future.

While in this reality, teaching and learning are still ongoing and cannot be stopped. New problems might occur in each new class. In this case, teachers should improve teaching proficiency by reflecting and finding solutions together with another teacher of the same subject. They could also consult the problems found with the school board. In this pandemic era, both teachers and the school's committee should work together to keep the learning and teaching process going as planned.

Prof. Yilin Sun from South Seattle College, USA stated in the Language Teacher Training and Education (LTTE) International Conference 2020 on the 25th of September, 2020 with the topic "Post-Pandemic Language Pedagogy: Perspectives and Directions' webinar" that whatever happens, teachers need to keep going and face these problems during the pandemic. Prof. Yilin Sun called it **resilience**. Resilience is where the teacher "bounces back" from the difficulties s/he faced and encourages his/her students to adapt with what happened to give the students hope and change them into agility where they learn from the problems they face and adapt to the environment. Particularly, Entesari et al. (2020) proved that resiliency happens significantly with experienced teachers, but this does not mean that novice teachers like the participant should give up. Instead, novice teachers should be able to learn from what happened and build their resiliency.

4 CONCLUSION

Based on the discussion above, the problems that emerge during online speaking class in this school is that the school's committee, teachers, and students are not yet ready to use the online platform. The school needs to rearrange its management and the facility both for teachers and students. The school's committee should also allow teachers to improve themselves by holding workshops every once in a while and consider hiring more teachers to lessen current teachers' workloads. If the school and teachers could provide a fun and engaging teaching and learning approach, it would allow students to perform better and be motivated to participate more in class.

As for the teacher, it is best for them to improve themselves by reflecting on their own teaching. To find solutions to the problems they found, they could discuss the problems whether with school or another teacher of the same subject. Teachers could also find literature regarding the problems. For problems regarding online teaching, there are many insightful webinars, and online workshops and teachers could look for the recorded webinars on YouTube. Also, teachers should encourage their students and motivate them, despite the inconvenient online teaching.

This study is a preliminary study that is far from perfect. The problems that arise during online speaking class in the COVID-19 pandemic era are different for each school and its elements. It is recommended to take further study with longer time and better instruments. Hopefully, this study would raise teachers' awareness regarding teacher professional development as it is important, especially in this pandemic era. Also, either the school committee or the teacher themselves should not take these issues for granted. It is important to reconsider policies and management in order to improve the quality of education for Indonesia's future generations.

REFERENCES

Alemi, M., & Tajeddin, Z. (2020). Reflection and good language teachers. *Lessons from good language teachers*, 41–53.

Ataiefar, F., & Sadighi, F. (2017). Lowering foreign language anxiety through technology: A case of Iranian EFL sophomore students. *English Literature and Language Review, 3*(4), 23–34.

Dewey, John. (1933). How we think. Madison, *WI: University of Wisconsin Press.*

Entesari, E., Yousefi, M. H., & Eslami, H. (2020). A mixed-method study of Iranian EFL teachers' achieving resiliency: implications for teacher development. *Asian-Pacific Journal of Second and Foreign Language Education*, 5(1), 1–20.

Moghaddam, R. G., Davoudi, M., Adel, S. M. R., & Amirian, S. M. R. (2019). Reflective Teaching Through Journal Writing: a Study on EFL Teachers' Reflection-for-Action, Reflection-in-Action, and Reflection-on-Action. *English Teaching & Learning*, 1–20.

Knapp, N. F. (2018). Increasing interaction in a flipped online classroom through video conferencing. *TechTrends*, 62(6), 618–624.

Lin, C. H., & Warschauer, M. (2015). Online foreign language education: What are the proficiency outcomes?. *The Modern Language Journal, 99*(2), 394–397.

Marull, C., & Kumar, S. (2020). Authentic Language Learning through Telecollaboration in Online Courses. *TechTrends*, 1–8.

McCoy, S., & Lynam, A. M. (2020). Video-based self-reflection among pre-service teachers in Ireland: A qualitative study. *Education and Information Technologies*, 1–24.

Nurkamto, J., & Sarosa, T. Engaging EFL Teachers in Reflective Practice as A Way to Pursue Sustained Professional Development. *International Journal of Pedagogy and Teacher Education, 4*(1), 45–58.

Peeters, W. (2018). Applying the networking power of Web 2.0 to the foreign language classroom: A taxonomy of the online peer interaction process. *Computer Assisted Language Learning, 31*(8), 905–927.

Prenger, Rilana; Poortman, Cindy L; and Handelzalt, Adam. (2017). Factors influencing teachers' professional development in networked professional learning community. *Teaching and Teacher Education*, 68, 77–90.

Rakes, G. C., & Dunn, K. E. (2015). Teaching online: Discovering teacher concerns. *Journal of Research on Technology in Education, 47*(4), 229–241.

Rotermund, S., DeRoche, J., & Ottem, R. (2017). Teacher Professional Development by Selected Teacher and School Characteristics: 2011-12. Stats in Brief. NCES 2017-200. *National Center for Education Statistics*.

Wang, C. (2020). Employing blended learning to enhance learners' English conversation: A preliminary study of teaching with Hitutor. *Education and Information Technologies*, 1–19.

Wong, M. S. (2011). Fifty ways to develop professionally: What language educators need to succeed. *Language Education in Asia, 2*(1), 142–155.

Post Pandemic L2 Pedagogy – Adi Putra & Arifah Drajati (Eds)
© 2021 Taylor & Francis Group, London, ISBN 978-1-032-05807-8

EFL teacher professional development in the pandemic era of COVID-19

Nurti Rahayu & Rina Suprina
Trisakti School of Tourism, Indonesia

ABSTRACT: This study investigates EFL teachers' effective professional development (TPD) during the pandemic. This study intends to answer whether or not EFL teachers can undergo TPD during the pandemic. It seeks to clarify how the pandemic has a positive effect on EFL teachers' TPD. It focuses on the EFL TPD in the tertiary education context. This research is descriptive qualitative, and the primary data is taken from questionnaires to the teachers. The findings reveal how EFL teachers improve their professional development during the pandemic and what professional development area they could manage. The result can be used to design the best TPD for EFL teachers, significantly to enhance educational technology in teaching and learning.

Keywords: EFL, EFL teachers, COVID-19, survey, teacher professional development

1 INTRODUCTION

Coronavirus 2019 (COVID-19) presents the most significant disruption in both international and national education. Fifty-three countries have to set up wide closures for their in-class education (UNESCO 2020). This pandemic resulted in short- and long-term impacts on Indonesian education (Syah 2020). The Circular of the Ministry of Education and Culture (Kemendikbud) of the Directorate of Higher Education No. 1 of 2020 aims to prevent the transmission of infection in higher education institutions by executing distance learning and instructing students to learn from their respective homes. The shift from conventional teaching to online learning presents some problems. Some of the problems are inadequate facilities and infrastructure, technological support equipment, limited and unstable internet networks, education budget to support online learning, and students' and teachers' lack of ICT competency (Syah 2020).

Teachers' professional development (TPD) is highly crucial for various reasons. First, TPD is associated with teacher satisfaction and retention (Schuck et al. 2011). In addition, teachers remain in their profession where they are assisted and educated in their practice and are always provided with ongoing advancement opportunities. (Gaikhorst et al. 2017). TPD is also a systematic attempt to contribute to a better in teaching practice, attitudes and beliefs and learning student outcomes. (Maher & Prescott 2017). Recent studies on TPD have focused on the use of online learning community (Lindberg & Olofsson 2009; Sari 2012; Powell & Bodur 2019), beginning and experienced teachers' PD (Coenders & Verhoef 2019), clusters of teachers who undergo TPD (Baecher & Chung 2020), informal PD with Facebook (Patahuddin & Logan 2019) and digital technologies (Fernandes et al. 2020; Nazari & Xodabande 2020).

Although considerable research has been explored in this field, less attention has been paid to investigating how EFL teachers in Indonesia undergo their TPD during pandemics. This particular interest is of great importance to confirm the ways teachers improve their professional development to conduct online teaching during the pandemic in tertiary education. Of course, the frequent webinars conducted for free serve as good opportunities for teachers to improve their professionalism. The chance for professional development (PD) in a pandemic is crucial to be investigated to

 DOI 10.1201/9781003199267-7

see how the teachers could spend their work-from-home with valuable professional development. Therefore, this research seeks to investigate how teachers benefit a lot from the teach-from-home in the pandemic; whether or not they can improve their PD. Given the significance, several themes on EFL teachers' PD, such as the kinds, goals, and areas of improvement, are elaborated.

2 LITERATURE REVIEW

The value of professional development (PD) on-campus performance is uncontested. Some of the research was taken from Hargreaves (1994) and Bolam (2000) as cited on (Fraser et al. 2007). However, the concept of professional development is still unclear. Hoban (2002), as cited by (Fraser et al. 2007), differentiated professional learning from professional development. Friedman et al. (2000), as cited in (Fraser et al. 2007) stated some professional development that is evident from the available literature. These include continuous learning for professionals; a means of self-improvement; a means for individual professionals to make sure a degree of control and protection in the often perilous workplace environment; a way of securing to the public that professionals are indeed up-to-date, expected to give the fast technology improvement; a means by which professional organizations can authenticate that the standard is in place (Fraser et al. 2007).

Therefore, much of the information was quickly forgotten, and the workshops are less practical. This development, then, needs to be re-conceptualized to align with their needs (Hunzicker 2011) to make an effective professional development program. Such program is referred to as structured professional learning that results in teacher knowledge and practices and improved student learning outcomes (Darling-Hammond et al. 2017). The phrase "professional learning" refers to a product of both externally provided and job-based activities that enhance the knowledge of teachers and help them improve their teaching practice in support of student learning. For this reason, the formal PD is a subset of the range of experiences that can lead to professional learning. Effective professional development should be as relevant as it is job-embedded, making it relevant and authentic. It is then instruction-oriented, involving the study and application of content and pedagogy, highlighting student learning outcomes and collaborative learning, engaging teachers in both active and interactive learning. Last but not least, a combination of contact hours, duration and consistency is ongoing. (Hunzicker 2011)

There is another term called continuous professional development (CPD) (Harland & Kinder 2014). The objective of the CPD is to make contributions to collective efforts to "to get good at change" through the CPD or In-Service Education and Training (INSET) of teachers by implementing the 'planned' change. Such proposed changes could be assessed as resulting in a number of outcomes, such as (1) material and provisional outcomes are physical resources that result from participation in INSET activities (e.g. worksheets, equipment, handbooks and time), (2) Informational outcomes, (3) New awareness, and (4) Value congruence outcomes (Harland & Kinder 2014). The results are mostly seen from a positive and substantial impact on teaching practice, which usually requires other intermediate outcomes, such as motivation and new knowledge and skills. In other words, CPD can also be seen as a means of introducing or enhancing knowledge, skills and attitudes by applying different models (Kennedy 2005).

To portray the EFL TPD in tertiary education in Indonesia, the researcher adopts some theoretical perspectives (Richards & Farrel 2005). The theories are selected as they often serve as a lens for the inquiry or generated during the study (Creswell 2014). They are in line with the present condition of EFL TPD in the society under study. They classify the types of PD activities into individual (self-monitoring, journal writing, critical incidents, teaching portfolios, action research), one-to-one (peer coaching, peer observation, critical friendships, action research, critical incidents, team teaching), group-based (case studies, action research, journal writing, teacher support groups), and institutional (workshops, action research, teacher support groups). Besides, the goals of PD are stated as to (1) become better informed about the field, (2) to learn more about learning strategies, (3) to develop more effective assessments, (4) to improve teaching, (5) to develop a better understanding of English grammar, and (6) to plan and evaluate a language course. The last aspect is the area of improvement can be seen from the cognitive strategies, improved knowledge, creativity and problem-solving, collaborative skills, intellectual-openness, work-ethic responsibility, and self-efficacy, self-regulation, and mental and physical health (Richards & Farrel 2005).

3 RESEARCH METHOD

Thirty-five EFL teachers responded to the online questionnaires distributed in EFL teachers WA Groups. The participants were recruited for the present study for two main criteria. First, they belong to the EFL WAG online community. Second, they teach various tertiary education in Jakarta and its nearby area. The majority of the participants were female, as much as 64%. The age group was 31-40 years old with 53%, and 25% comes from those 41–50 years old. As much as 44% have been teaching less than a year; the other 42% have 10-15 years of teaching experience. Most of the participants had a master's degree (S2) with 92%, S3 with 6%, and the rest from S1. The teaching locations were from various provinces in Java.

This study uses a case study to investigate a contemporary phenomenon within its real-life context (Yin 2003). The data collection is performed by distributing online questionnaires in online community networking. Before the items were distributed, they had been reviewed by some colleagues for valuable feedback for item clarity (Dörnyei 2003). Convenient sampling was used due to its easiness of access and network available in EFL teachers WhatApps Group. The questionnaire consisted of several parts (1) demographic data, (2) types of PD, (3) the goals of PD and (3) the area that has improved during PD. The types of PD ranges from individual, one-to-one, group-based, and institution PD (Richards & Farrel 2005). The participants were requested to answer how often they conducted each type of activity. The other parts were asking about the goals of TPD and areas of improvement with Likert scales. The data were analyzed for minimum, maximum, mean, and standard deviation.

4 FINDINGS AND DISCUSSION

The results present the research findings in three areas: types of PD, the goals of PD, and area of improvement.

4.1 *Types of PD*

A list of personal, one-on-one, group and institutional PD was listed with five Likert Scale with the degree of frequency stating (1) never to (5) always. Range of mean is used as the basis of analysis. The result is presented in this table below.

Table 1. Types of TPD.

Types of PD	Min	Max	Mean	SD
Webinar	1	5	3.06	1.027
Online Workshop	1	5	2.97	1.014
Writing Journal	1	5	2.77	0.973
Teaching Reflection	1	4	2.74	0.78
Critical incident	1	4	2.6	0.881
Team teaching	1	4	2.51	1.067
Teaching portfolios	1	4	2.49	0.818
T_ support group	1	4	2.46	1.01
Case studies	1	4	2.34	0.968
Peer coaching	1	5	2.31	1.105
Action Research	1	5	2.29	0.987
Peer observation	1	4	2.29	0.987
Critical Friendship	1	4	2.26	1.01

The items of PD were classified as personal, one on one, group, and institutional PD. The table shows the webinars as the highest mean, with 3.06 showing the highest frequency. However, it is classified as "sometimes," some participants' rate "5," showing that they always participate in the webinars. The same thing applies to online workshops and writing journals. It means that the highest means of PD is institution PD, i.e., webinars and online workshops.

The lowest mean belongs to action research, peer observation and critical friendship. The WAG discussion revealed that most EFL teachers had vague ideas on what critical friendship is. This PD has not gained much popularity compared with others. Critical friendship occurs when two individuals or more are involved in mutual professional development. They believe that they can enhance their teaching practice by doing discussions, reflections and conducting a research project together. (Golby & Appleby 1995).

The idea of being a critical friend has been around since the 1960s with the idea that "critical friends" are people suggested by Stenhouse (1975) to take up pro-active roles by establishing academic partnership in the whole project (Kember et al. 1997). These roles should be consistently applied to self-study phases (Schuck & Russell 2005). In this process, dialogue and learning occurs in the context of the leadership of learning (Swaffield 2008). Regarding such roles, EFL teachers seem to have such kinds of reluctance to provide guidance and assistance with their peers, which are worded as professional indifference, based on various factors (Baskerville & Goldblatt 2009). The details above confirmed how teachers claimed PD as merely an institution program that they have to join. There is less exploration of other personal and one-on-one professional development to enhance their teaching quality.

4.2 *The goals of PD*

Some statements were used to investigate the teachers' goals of their PD. The responses were captured in four Likert Scales with (1) strongly disagree, and (4) strongly agree. The responses were analyzed in terms of the agreement percentage, the percentage of responses with strongly agree and agree, minimum, maximum score, mean and standard deviation.

Table 2. The goals of PD.

PD Goals	% Agreement	Min	Max	Mean	SD
Better informed	91.4	1	4	3.49	0.724
Develop effective assessment	94.3	1	4	3.20	0.719
Teaching improvement	91.4	1	4	3.14	0.81
Plan & evaluate EFL course	94.3	1	4	3.20	0.719
Learn teaching strategies	91.4	1	4	3.09	0.781
Understand English grammar	91.4	1	4	3.14	0.733

The goals of PD were stated in six statements: to be better informed, to develop effective assessment, to improve teaching, to plan and evaluate EFL courses, to learn teaching strategies, and to understand English grammar. It can be seen from the percentage of "strongly agree" and "agree" responses were more than 90%. The most dominant goals stated were to plan and evaluate EFL courses and to develop effective assessment with 94.3%. The other goals were 91.4%.

The findings are interesting regarding the need to plan, evaluate, and assess online classes in the pandemic. Results from the WAG discussion also confirmed that there was an urgent need to execute the EFL classes online. Some teachers mentioned that they previously learned about online and blended learning from reference books and research articles, but most of them had no previous experience conducting online classes. Due to the pandemic, teachers have to execute the classes suddenly with a lack of preparation. In this case, teachers' priority for PD was to prepare for the classes, starting with the lesson plan with synchronous and asynchronous classes, transform the offline lesson plan into the online lesson plan, adjust the teaching sequence, select the materials, evaluate the teaching-learning process. They finally have to prepare for an adequate assessment covering the curriculum's knowledge, skills and attitude.

Other goals such as getting better informed, teaching improvement, learning teaching strategies and improving grammar also gain high agreement with 91.4%. It can be inferred that the pandemic encourages the teachers to experience new ways of teaching out of their comfort zone. In order to fulfill their responsibility, they need to go through some professional development. The goals of PD can also be seen from the mean score. In line with the percentage, most EFL teachers agree with the questionnaires' goals, starting from being better informed in subject knowledge with 3.49 to learning teaching strategies with 3.09. This fact is in line with Klinzing and Tisher (1993) and Lewin (1990b) as cited by (Thair & Treagust 2003). They confirmed that the education quality is directly associated with teachers' professional development in the aspects of methodology of teaching and subject knowledge. These research findings add to the literature of teachers' PD in Indonesia. Not much research was conducted to investigate the EFL in-service teachers' PD in higher education. Most of the research focused on TPD for pre-service teachers from teaching in another subject (Hadi 2002), TPD initiatives in the education department in universities, and government initiative program through UKG (Thair & Treagust 2003)

4.3 *Areas of improvement*

The last research question is about the area of improvement of PD during pandemic since the statements were specified to the last six months, which were related to the pandemic outbreak. The results were presented as the following.

Table 3. Area of improvement.

Improvement	% Agreement	Min	Max	Mean	SD
Processing Cognitive strategies	83	1	4	2.94	0.639
Improved knowledge	97	1	4	3.11	0.530
Creativity & Problem solving	89	1	4	2.97	0.664
Collaborative group skill for	77	1	4	2.87	0.758
Intellectual openness	91	1	4	3.06	0.591
Work ethic responsibility	91	1	4	3.03	0.568
Self-efficacy, self regulation, mental, physical health	77	1	4	2.91	0.781

The table confirmed the area of improvement, which was formulated in eight (8) statements. The research participants mostly agree that their PD has improved their knowledge of related subjects. The fact was proven with 97% agreement and a 3.11 mean score. Other affected areas were intellectual openness, and work ethic responsibility with 91% agreement, 3.06 and 3.03 mean.

The further discussion confirmed that intellectual openness was partly associated with the acceptance of ICT use in online teaching and blended learning methods. Long before the pandemic, the use of ICT in teaching and blended learning gained both supports and resistance from teachers. Those teachers believe in conventional face-to-face teaching as the best teaching method. At some points, they have not got the whole picture on how to use ICT in online teaching with an insufficient internet connection and lack of infrastructures. Long before the pandemic, Nugraha (2009) as cited by (Widodo & Riandi 2013) proved the benefits of ICT-based learning such as improving students' understanding and creative thinking and motivation Suhendi (2009) as cited by (Widodo & Riandi 2013) and learning outcomes; Higgins and Moseley (2001) as cited by (Widodo & Riandi 2013).

The findings confirmed that the EFL teachers make the most of their time by attending the webinar and workshops during the pandemic. These two activities were more popular than other TPD activities, such as peer observation and critical friendship. Most EFL teachers undergo TPD to be better informed and to develop an adequate assessment. The last goal is prevalent as teachers often ask this topic in the webinar Q&A. EFL teachers have not understood an online assessment for English subjects. It is also worth noting that their professional development results in knowledge improvement and intellectual openness, two essential aspects of personal development. Given those findings, teachers' professional development in an Indonesian context is always unique phenomena to discuss, and it goes through a series of models which affect the classroom practices, later change the students' learning outcome, and later change the teachers' beliefs and attitude (Guskey 2002). The pandemic provides unique phenomena when teachers have to learn, adopt, and adapt new insights to online classroom practices.

5 CONCLUSION AND SUGGESTION

The current study set out to report the types, goals, and areas of EFL TPD improvement. This study showed how EFL teachers made the most of their time to attend the webinar, online workshops, and other activities. The pandemic forced them to shift from the classroom into online teaching, and teachers need to be more knowledgeable and open-minded. This research seeks to answer three initial research questions. First, EFL teachers mostly participate in institutional PD such as webinars and online workshops during the pandemic. The lucrative webinars and online workshops available for free motivate the teachers to take part and gain some benefits. The second finding confirmed that PD's dominant goals were to plan, evaluate and develop effective assessments for EFL courses in the online context. The last finding revealed that most EFL teachers perceived that their PD has improved their knowledge on their subject matter, intellectual openness and work ethic responsibility. Despite the perceived improvement area, the research participants think their PD has the least impact on collaborative group skills, self-efficacy, self-regulation and mental and physical health.

However, those findings did not reflect the EFL teachers' overall perceptions in a more significant population since this study only utilized small participants. Despite these limitations, this study revealed the initial depiction of EFL teachers' efforts for self-improvement to keep up with the challenges of online learning. At the same time, it expanded the scope of discussion involving the preferred activities, goals, and the outcome. This study hopefully can serve as a modest step toward more advanced research on EFL TPD, which can be used as an initial model to develop EFL TPD for the near future. Further researchers could explore the relationships of some variables involved in TPD.

REFERENCES

Baecher, L., & Chung, S. (2020). Transformative professional development for in-service teachers through international service-learning. *Teacher Development, 24*(1), 33–51.

Baskerville, D., & Goldblatt, H. (2009). Learning to be a critical friend: From professional indifference through challenge to unguarded conversations. *Cambridge Journal of Education, 39*(2), 205–221.

Coenders, F., & Verhoef, N. (2019). Lesson Study: professional development (PD) for beginning and experienced teachers. *Professional Development in Education, 45*(2), 217–230.

Coolahan, J. (2002). Teacher Education and the Teaching Career in an Era of Lifelong Learning. *OECD Education Working Papers, 2*, 39.

Creswell, J. W. (2014). *Research Design. Qualitative, Quantitative, and Mixed Methods* (4th edition). Sage Publications, Inc.

Darling-hammond, L., Hyler, M. E., & Gardner, M. (2017). Effective Teacher Professional Development. *Learning Policy Institute, June*.

Dörnyei, Z. (2003). Questionnaires in Second Language. In *Mahwah: Lawrence ErlbaumAssociates*.

Fernandes, G. W. R., Rodrigues, A. M., & Ferreira, C. A. (2020). Professional Development and Use of Digital Technologies by Science Teachers: a Review of Theoretical Frameworks. *Research in Science Education, 50*(2), 673–708. https://doi.org/10.1007/s11165-018-9707-x

Fraser, C., Kennedy, A., Reid, L., & McKinney, S. (2007). Teachers' continuing professional development: Contested concepts, understandings and models. *Journal of In-Service Education, 33*(2), 153–169.

Gaikhorst, L., Beishuizen, J. J. J., Zijlstra, B. J. H., & Volman, M. L. L. (2017). The sustainability of a teacher professional development programme for beginning urban teachers. *Cambridge Journal of Education, 47*(1), 135–154.

Golby, M., & Appleby, R. (1995). Reflective Practice through Critical Friendship: Some possibilities. *Cambridge Journal of Education, 25*(2), 149–160.

Guskey, T. R. (2002). Professional development and teacher change. *Teachers and Teaching: Theory and Practice, 8*(January 2013), 381.

Hadi, S. (2002). *Effective teacher professional development for the implementation of realistic mathematics education in Indonesia* [University of Twente].

Harland, J., & Kinder, K. (2014). Teachers' Continuing Professional Development: Framing a model of outcomes. *Professional Development in Education, 40*(4), 669–682. https://doi.org/10.1080/19415257.2014.952094

Hunzicker, J. (2011). Effective professional development for teachers: A checklist. *Professional Development in Education, 37*(2), 177–179.

Kember, D., Ha, T. S., Lam, B. H., Lee, A., Ng, S., Yan, L., & Yum, J. C. K. (1997). The diverse role of the critical friend in supporting educational action research projects. *Educational Action Research, 5*(3), 463–481.

Kennedy, A. (2005). Models of Continuing professional development: a framework for analysis. *Journal of In-Service Education, 31*(2).

Lindberg, J. O., & Olofsson, A. D. (2009). Online learning communities and teacher professional development: Methods for improved education delivery. In *Online Learning Communities and Teacher Professional Development: Methods for Improved Education Delivery*.

Maher, D., & Prescott, A. (2017). Professional development for rural and remote teachers using video conferencing. *Asia-Pacific Journal of Teacher Education, 45*(5), 520–538.

Nazari, M., & Xodabande, I. (2020). L2 teachers' mobile-related beliefs and practices: contributions of a professional development initiative. *Computer Assisted Language Learning, 0*(0), 1–30.

Patahuddin, S. M., & Logan, T. (2019). Facebook as a mechanism for informal teacher professional learning in Indonesia. *Teacher Development, 23*(1), 101–120.

PDDikti. (2017). *Sebaran Dosen di Indonesia*.

Powell, C. G., & Bodur, Y. (2019). Teachers' perceptions of an online professional development experience: Implications for a design and implementation framework. *Teaching and Teacher Education, 77*, 19–30. https://doi.org/10.1016/j.tate.2018.09.004

Richards, J.C & Farrel, T. (2005). *Professional Development for Language Teachers* (First). Cambridge University Press. https://doi.org/10.1017/CBO9781107415324.004

Ristekdikti. (2018). *Statistik Pendidikan Tinggi 2018*. https://pddikti.ristekdikti.go.id/asset/data/publikasi/Statistik Pendidikan Tinggi Indonesia 2018.pdf

Sari, E. R. (2012). Online learning community: A case study of teacher professional development in Indonesia. *Intercultural Education, 23*(1), 63–72.

Schuck, S., Aubusson, P., Buchanan, J., Prescott, A., Louviere, J., & Burke, P. (2011). *Retaining effective early career teachers in NSW schools* (Issue December).

Schuck, S., & Russell, T. (2005). Self-Study, Critical Friendship, and the Complexities of Teacher Education. *Studying Teacher Education, 1*(2), 107–121.

Swaffield, S. (2008). Critical friendship, dialogue and learning, in the context of Leadership for Learning. *School Leadership and Management, 28*(4), 323–336.

Syah, R. H. (2020). Dampak Covid-19 pada Pendidikan di Indonesia: Sekolah, Keterampilan, dan Proses Pembelajaran. *SALAM: Jurnal Sosial Dan Budaya Syar-I, 7*(5).

Thair, M., & Treagust, D. F. (2003). A brief history of a science teacher professional development initiative in Indonesia and the implications for centralised teacher development. *International Journal of Educational Development, 23*(2), 201–213.

Widodo, A., & Riandi. (2013). Dual-mode teacher professional development: challenges and re-visioning future TPD in Indonesia. *Teacher Development, 17*(3), 380–392.

Yin, R. K. (2003). *Case Study Research Design and Methods* (pp. 1–181).

Post Pandemic L2 Pedagogy – Adi Putra & Arifah Drajati (Eds)
© 2021 Taylor & Francis Group, London, ISBN 978-1-032-05807-8

Investigating English teachers' online learning engagement: A case study during COVID-19 pandemic

Pratiwi Amelia, Dwi Rukmini, Januarius Mujiyanto & Dwi Anggani Linggar Bharati
Universitas Negeri Semarang, Indonesia

ABSTRACT: This study investigates English teachers' online learning engagement in teaching English situated in a pandemic COVID-19. Thus, a case study involving four junior secondary English teachers in Indonesia was conducted. The instrument used for this investigation was interviews and observations. The TPACK framework and online learning models were used as a theoretical base in this study. The research findings revealed that the teachers had tailored online learning and attempted to adjust a new practice. As a result, they experienced different challenges when coping up technological, pedagogical, and content challenges. This study underlines the importance of teachers' self-awareness, self-motivation, and independence to education continuity, to support sustained teacher professional development. This study implies the importance of social support and school assistance to strengthen the teachers' knowledge, particularly in teaching English at junior secondary school.

Keywords: COVID-19, English teachers, online learning, TPACK

1 INTRODUCTION

Due to the rapidly increasing cases of COVID-19 in Indonesia since March 2020, the government of Indonesia through The Ministry of Education and Culture agreed to suspend all face-to-face learning at schools. Thus, the government obligated the schools to transform their teaching into online learning as part of the pandemic control decision during the outbreak. As the main alternative to maintain the pedagogical change during the pandemic, the government offered the program of "*Belajar dari Rumah*" or learning from home through online learning to continue education in a time of crisis during pandemic COVID-19. This program encompasses revisiting the pedagogical approach, modifying course objective, assessment approach, and content. Due to this policy, many schools implemented online learning suddenly to turn into online learning without any preparation. However, although the government has offered some alternatives to continue learning during COVID-19, it does not merely solve some problems in implementing online teaching. Some areas still have limited access and lack of teaching knowledge, particularly in the remote teaching (Lie et al. 2020). Besides, there are some problems such as lack of direct interaction (Sepulveda-Escobar & Morrison 2020), lack of practical experiences (Donitsa-Schmidt & Ramot 2020).

As the teachers are obligated to turn into online teaching and learning, they have to construct and reconstruct their knowledge to deliver online learning. Pedagogically, the teachers had learned how to design and implement learning, manage student learning, and conduct an assessment as they have practised so far in face-to-face learning. However, they have not yet gained in-depth knowledge about how to teach with online learning. Besides, in face-to-face mode, the teacher did not use much technology; textbooks are the most widely used. Ironically, during this pandemic, the teachers are expected to set online learning by making significant changes in their lesson.

Meanwhile, to do this, the teachers need sufficient knowledge of technology, pedagogy, or content in online learning settings. Since technological tools in online learning are a primary need to connect teacher-students or students-students in the distance, TPACK framework as the basic knowledge should be implemented. Mishra and Kohler (2019) mention that the technological, pedagogical, and content knowledge (TPACK) framework should be used as the primary consideration in integrating technology into a lesson. Relearned and readapted technology, pedagogical, and content is one of the aspects that are required during the suspension of face-to-face learning (Lie et al. 2020). The TPACK model for online learning should be implemented (Nasri et al. 2020). So, based on this empirical evidence, it can be said that online learning and TPACK are interrelated. That is why online learning models and TPACK were used as the theoretical base in this study.

2 CONTEXT AND EFL TEACHING IN INDONESIAN JUNIOR SECONDARY SCHOOL

This study investigates four junior secondary English teachers' online learning engagement while delivering online learning into an English lesson for junior school students aged 12–15 years old. To begin with, we would like to introduce context details of English education in Indonesian junior secondary teaching in Indonesia. English has been taught in Indonesian junior school as a foreign language since 1989 through Law Year 1989 Chapter IX Section 39 and Ministerial Regulation No 23 the Year 2006. English is a compulsory subject to teach at the secondary level. All schools have implemented English language learning in schools with an average of one-week lessons per meeting. However, in response to language teaching and COVID-19 pandemic in Indonesia, Lie et al. (2020) reported that the pandemic forced the English teachers to turn into online learning. Besides, based on observation, most English teachers in Indonesia are not familiar with online learning. Therefore, based on the current constraints, we are eager to investigate the English teachers' online learning engagement while delivering online teaching, which focuses on the remote area teaching in four regions in Indonesia. It is hoped that this study will give pieces of evidence that could provide more attention and intervention from education authorities to the success of online learning during COVID-19 pandemic. Therefore, the research questions of this study are developed as follow:

1) How did the junior secondary English teachers engage in online learning during the COVID-19 pandemic using TPACK and online learning model as a conceptual framework?
2) What technological, pedagogical, and content challenges did the junior secondary English teachers face while engaging in online learning during the COVID-19 pandemic?

3 LITERATURE REVIEW

3.1 *Concept of TPACK and online learning*

The sudden closing of educational institutions led to remote online learning where the technology to ensure learning continuity is using the most (Nasri et al. 2020; Lie et al. 2020). This study used the TPACK framework proposed by Mishra and Koehler (2006) to evaluate teachers' experience in online learning and the online learning model developed by Anderson (2008) to investigate the teachers' learning experience. Mishra and Koehler (2006) introduced the framework of TPACK to help the teachers to construct their knowledge on ICT integration and provide teachers with systematic design thinking that expands the idea of Shullman's (1986) pedagogical content knowledge (PCK). la Velle and Leask (2019) stated that PCK is one factor that contributes significantly to effective instruction. Since technological tools have been widely used over the years, technological knowledge should be incorporated into the PCK framework (Hofer & Grandgenett 2012).

Mishra (2019) describes TPACK as the fundamental knowledge drawn from a unique integration of technology, pedagogy, and content to highlight specific subjects and contexts to fulfil learners' varied backgrounds. Mishra and Koehler (2006) presented a systematic view of the knowledge required in technology integration. It consists of the knowledge: 1) Content Knowledge (CK) refers to the specific subject matter or content knowledge, 2) Pedagogical Knowledge (PK) refers to teaching approach, curriculum, students' background, and 3) Technological Knowledge (TK) defines as the knowledge of using various technology for educational purposes. The other four domains in TPACK framework are the interplay between CK, PK, and TK including knowledge to blend a practical pedagogical approach with the content (PCK), the knowledge to integrate appropriate technological tools with pedagogical practice (TPK), the knowledge to integrate technological knowledge with the specific content (TCK), and the knowledge to integrate technological, pedagogical content knowledge (TPACK). Recently, Mishra (2019) presented additional domains on the TPACK framework, namely "contextual knowledge" which refers to the contextual factors such as policy, teachers' awareness, and technological availability.

Furthermore, Anderson (2008) presented six forms of interaction for effective online and remote teaching and learning. The interaction in online learning encompasses student-student interaction, student-teacher interaction, student-content interaction, teacher-teacher interaction, teacher-content interaction, content-content interaction. There are two kinds of interaction that is aligned with this study; it encompasses 1) student-content interaction, which refers to students' interaction toward the content in online learning, and 2) student-teacher interaction, which refers to the communication, and the interaction between students and teachers in online learning. In this study, the TPACK framework and online learning models were employed. Both models are complementary, and it will ultimately provide insight into the intersection between theory and reality, especially for the implementation of online teaching in teaching English for the junior secondary classroom during COVID-19 pandemic.

4 RESEARCH METHOD

This study used a case study design (Stake 1995) to explore junior secondary English teachers' online learning engagement. This study is part of systematic research that was carried out in term two the Year 2020. This study was conducted in four regions in a remote area in Indonesia: Bangka Belitung, East Java, South East Sulawesi, and Maluku. The recruitment of participants in this study was considered through two main reasons. First, the participants have a strong willingness to provide information related to their online learning engagement. Second, the participants conduct online teaching, both synchronous and asynchronous during COVID-19 pandemic.

The participants in this study were Rose, Audrey, Ahmed and Maria (pseudonym); they work in a junior secondary school in Indonesia. No participants in this study have ever done full online learning before pandemic COVID-19. In collecting the data, we use semi-structured interviews to investigate teachers' online learning engagements while teaching English during COVID-19. In this study's recruitment, we first contacted the participants through *WhatsApp* to ask their willingness to participate in this research. The consent form and invitation letter were also addressed to all participants to explain the research procedure in detail written in Bahasa Indonesia. After the participants agreed to participate, we guided the participants to sign off on all items on the consent form to ensure all data were kept confidential. The interview was conducted by using *Zoom* to meet the participants virtually. We asked the consent from participants to record the conversation during interviews. The interview was conducted based on the participants' agreement and the participants' time availability. The data drawn from the interview is classified in the thematic analysis to identify the findings. The data were also drawn from observations. We ask consent from participants to observe their online learning practice.

5 FINDINGS AND DISCUSSION

5.1 *Junior secondary English teachers' online learning engagement*

This study investigates how junior secondary English teachers make engagement in their online learning during COVID-19 pandemic. We observed the teachers' online teaching practice for six weeks to analyse their online learning engagement. The rubric of online learning engagement based on TPACK (Lie et al. 2020) was employed. It found that the implementation of online learning during COVID-19 brings up new experiences to all participants. They presented a varied level of online learning engagements in their classroom, as presented in table 1.

Table 1. The presentation of participants' online learning engagement level.

Levels of online learning engagement	Regions				
	Bangka Belitung	East Java	South East Sulawesi	Maluku	Total
None/Almost None					0
Rudimentary					0
Basic		Ahmed		Maria	2
Intermediate	Audrey		Rose		2
Advanced					0

Based on Table 1, it can be seen that there are two participants in the level of basic (Ahmed and Maria), and the rest are in the level of intermediate (Audrey and Rose), and none of them is in the level of none/almost none, rudimentary, and advanced. It means that online learning was implemented in their classroom following the recommendations set by the government. However, the ways they engage in online learning were not the same. For example, based on interviews Maria conveyed that she used *WhatsApp* and *PowerPoint* for technological tools, discussions for pedagogical approach, and employed local legend/tales for the content of materials to attract the students' attention. However, only a few students in real teaching practice gave responses because they have low motivation to study, low confidence to speak, lack of English competence, lack of technological knowledge, and lack of technological availability. Therefore, interaction and engagement were very minimal.

Meanwhile, in Audrey' classroom, she conveyed that she selected various technological tools (e.g. *Zoom, WhatsApp, Powerpoint, ClassDojo,* and *Filmorago*) to enhance learning. She used project-based as the approach. She was also selected the local legend/tales for the content. However, the students in Audrey' class have a strong motivation to learn English. They were confident to speak English, even though sometimes they still needed the teacher's help to check their utterances. Besides, most of the students do not have problems with using technologies. As a result, the students participated and engaged actively in the class.

It can be said, the enactment of online learning through technological tools has existed in each teachers' teaching practice through the enactment of various technological and pedagogical approaches, but their online learning engagement is varied. As stated by Anderson (2008), the interaction between the teachers-students, students-content interactions are one of the factors that influence the effectiveness of online learning engagement as experienced by Rose as follow:

For example, Rose reported:

> *"I have learned some technological tools and applications to enrich English lessons in my class. I have tried to change the way of interaction with the students by using some applications, such as Zoom, WhatsApp, and Google Classroom. I used Class Dojo and Screen O' Matic to attract the students' attention. Some of the students engaged actively; some of them were not fully engaged."*

Although the teachers had engaged and employed varied technological and pedagogical approaches to adapt to online learning, it does not imply that they engage with the students or make them interact with teachers or peers during the class. Nasri et al. (2020) stated that some teachers faced challenges in adapting social interaction during online classes because they might have different needs and different technological experiences (Nasri et al. 2020). In this study, online learning teachers have applied technology, pedagogy, and content, but it has not yet fully developed students' English skills. We suggest that strengthening teacher knowledge extensively for online learning is needed so that teachers and students can be more engaged.

5.2 *Challenges of online learning faced by junior secondary English teachers*

Pandemic COVID-19 pandemic has created unique challenges for teachers, including problems caused by emergency conversion to online language teaching (MacIntyre et al. 2020). In this study, the junior secondary English teachers in junior secondary classrooms expressed their challenges with a different perspective and experience. We categorised their challenges into three issues: technological, pedagogical, and content challenges. The evidence shows that the first challenges faced by the teachers were technological challenges. It can be said that this is the significant challenge faced by the teachers. For example, as Ahmed articulated, when he tried to prepare the lessons, the account's first considerations are the technological considerations (e.q. accessibility, availability, and affordability). Besides, he articulated that sometimes he was nervous when using some technological features due to lack of confidence and lack of knowledge to use it appropriately.

For example, Ahmed reported:

> *"I felt nervous, and I was not so confident even though I have learned and prepared to use Zoom for video conferencing during online teaching in my class. However, sometimes I forget how to operate certain features on it. So sometimes I focus too much on how to solve technological problems, so the learning strategies and materials I have compiled are not well presented."*

Besides the technological challenges, the participants faced pedagogical challenges and content challenges. Richard (2017) proposed that pedagogical knowledge refers to teaching knowledge, including the content knowledge and teaching techniques and approaches used by the teachers following the teachers' philosophies, views, principles, values and ideas. Pedagogical knowledge also encompasses the learners' knowledge, the curriculum, the teaching context and the teaching methods (Richards 2017). In this study, the teachers conveyed how difficult it was to foster student motivation during online learning, as experienced by Maria as follows:

Maria reported:

> *"The motivation and the participation of students during online learning in COVID-19 pandemic were shallow. To solve this problem, I conduct personal approaches by engaging students in face-to-face communication to find their depth problems. I also continuously make intensive communication with parents to keep in touch with the students' progression during online learning."*

As motivation and the students' engagement were shallow, Maria employed a personal approach to digging out the students' problems. This interaction refers to the idea proposed by Anderson (2008), which states that student-teachers interaction or student-student interaction is a requirement of distance education. Besides, this study found that not all students engage with the content or the teachers' materials. To overcome this, the teacher must be creative in choosing the appropriate material for students not to get bored. Project assignments can be an alternative to make students more active and creative independently by integrating the use of technology, such as digital project assignments, which can be useful to enhance the students' language skills, multimodal, and digital literacies (Jones & Chapman 2017; Lestariyana & Widodo 2018; Drajati et al. 2018; Churchill 2020).

6 CONCLUSION

Using technology to promote learning during school closure is a requirement (Lie et al. 2020). Thus, the teachers need to reshape their knowledge to fulfil various needs of learners. Awareness of the influential factors on online learning implementation will help solve challenges and provide successful learning opportunities. The suspension of face-to-face learning during COVID-19 pandemic has created some challenges. Teachers' motivation and awareness to improve their career growth are considered more positive than negative emotions such as confused, anxious, and stressed. This study reveals that the obligation to teach online during COVID-19 encouraged the participants to acquire new teaching directions. This finding implies the significance to acquire technological, pedagogical, and content knowledge to ensure learning continuity and equity (Nasri et al. 2020). Besides, resilience, eagerness, teachers' self-confidence, willingness to adapt, and social and school assistance will support the teachers' online learning engagement during COVID-19 pandemic.

REFERENCES

Anderson, T. (2008). *The theory and practice of online learning.* 2nd ed. Edmonton, AB, Canada: AU Press.

Churchill, N. (2020). Development of students' digital literacy skills through digital storytelling with mobile devices, *Educational Media International.* 57(3), 271–284.

Donitsa-Schmidt. S, & Ramot, R. (2020). Opportunities and challenges: teacher education in israel in the COVID-19 pandemic, *Journal of Education for Teaching*, 46(4), 586–595.

Drajati, N. A., Tan, L., Haryati, S., Rochsantiningsih, D., & Zainnuri, H. (2018). Investigating english language teachers in developing TPACK and multimodal literacy. *Indonesian Journal Of Applied Linguistics*, 7(3), 575–582.

Hofer, M., and N. Grandgenett. (2012). TPACK development in teacher education. *Journal of Research on Technology in Education*, 45 (1), 83–106.

Ia Velle, L., and M. Leask. (2019). *What do teachers do? In learning to teach in the secondary school: a companion to school experience*, edited by S. Capel, M. Leask, and S. Younie, 9–27. 8th ed. Abingdon: Routledge.

Jones, S., & Chapman, K. (2017). Telling stories: Engaging critical literacy through urban legends in an english secondary school. *English Teaching: Practice & Critique*, 16, 85–96.

Lestariyana, R. P. D., & Widodo, H. P. (2018). Engaging young learners of English with digital stories: Learning to mean. *Indonesian Journal of Applied Linguistics,* 8, 488–494.

Lie, A., Tamah, S. M., Gozali, I., Triwidayati, K. R., Utami, T. S. D., & Jemadi, F. (2020). Secondary school language teachers' online learning engagement during the COVID-19 pandemic in indonesia. *Journal of Information Technology Education: Research*, 19, 803–832.

Nasri, H., N, Husnin, H., Mahmud, D., N., S & Halim, L. (2020) Mitigating the COVID-19 pandemic: a snapshot from Malaysia into the coping strategies for pre-service teachers' education, *Journal of Education for Teaching*, 46 (4), 546–553.

MacIntyre, P. D., Gregersen, T., & Mercer, S. (2020). Language teachers' coping strategies during the COVID-19 conversion to online teaching: correlations with stress, wellbeing and negative emotions. *System*, 94, 102352.

Mishra, P., & Koehler, M. J. (2006). Technological pedagogical content knowledge: a framework for teacher knowledge. *Teachers College Record,* 108(6), 1017–1054.

Mishra, P., (2019). Considering contextual knowledge: the TPACK diagram gets an upgrade. *Journal of Digital Learning in Teacher Education*, 35 (2), 76–78.

Richards, J. C. (2017). Teaching english through english: Proficiency, Pedagogy and Performance. *RELC Journal*, 48(1), 7–30.

Sepulveda-Escobar, P., & Morrison. A. (2020). Online teaching placement during the COVID-19 pandemic in Chile: challenges and opportunities, *European Journal of Teacher Education*, 43 (4), 587–607.

Shullman, L. S. (1986). Those who understand: knowledge growth in teaching. *Educational Researcher*, 15(2), 4–14.

Stake, R. E. (1995). *The art of case study research.* Thousand Oaks: Sage Publications.

Post Pandemic L2 Pedagogy – Adi Putra & Arifah Drajati (Eds)
© 2021 Taylor & Francis Group, London, ISBN 978-1-032-05807-8

Multimodality in English learning for hard-of-hearing learners during the COVID-19 pandemic

Nur Arifah Drajati, Rizka Junhita & Bunga Ikasari
Universitas Sebelas Maret, Indonesia

ABSTRACT: Students with Special Education Needs (SEN) have the same needs as general students. However, due to the COVID pandemic outbreak, the learning process should be done at home. Teachers are obligated to use a multimodal approach that integrates all verbal and visual semiotic modes and ICT in the classroom. This phenomenon creates teachers that teach using technology. Thus, this study aimed at exploring the use of the digital literacy storybook practice made by teachers to teach their special education students who study at home. The two respondents consisted of teachers of hard-of-hearing students who are interviewed individually. The data gained from the interview was analyzed by thematic analysis based on themes that emerged from the codes and categories. The study's findings revealed that digital storybooks help teachers implement the practice for special education needs students. Also, a digital storybook for special education students allows students to be better assisted, which offers multiple multimodalities at the same time to support special students in learning English easily.

Keywords: digital storybook, narrative inquiry, online class, special education needs

1 INTRODUCTION

English is considered one of the most popular subjects taught at school. Students with special needs, such as hard-hearing (HH) students, autism students, or others also learned English at school (Oyler 2011). However, their ability to process the materials was quite different from students without special needs (Jones & Enriquez 2009). Some problems might occur in the learning process due to the complexities in the inclusive classroom. However, visually delivered materials can be one of the solutions (Bray et al. 2014). Special education needs some modes to support the learning process (Chong & Daud 2016). Some modes are facial expression, hand gesture and lip-reading. Modes are parts of multimodality. Kress (2010) described image, layout, gesture, objects and speech as part of modes. These modes can be used to support the learning process. In a previous study conducted by Dennis (2015), teachers take all the actions to create engagement in the learning process using digital with special needs students.

2 LITERATURE REVIEW

2.1 *Special education need students*

People with developmental disabilities had similar struggles (Biklen 2000). Struggles occurred because the students had special characteristics (autism, cerebral palsy, down syndrome, hard-hearing). However, students with particular characteristics are familiar with the use of technology (Williams 2006). The use of technology in special education requires a wide range of applications (Standford et al. 2010). Because of this phenomenon, literacy and semiotics are associated with

DOI 10.1201/9781003199267-9

new literacy studies. The increasing role of digital technologies for communication is one of the reasons (Mills 2010)

Rahamin (2004) stated that ICT had been used to support special education needs learners for many years. Thus, educators also need to have essential sensitivity and sound educational practice to ensure that students with particular characteristics can learn better (i.e., engage with the material and technology development). Besides that, inclusive teachers or parents who engage in innovative pedagogies related to digital literacies focus on learning experiences (Price-Dennis et al. 2015).

2.2 *Multimodality in digital storybook*

Digital storybooks support story-telling activities using multimedia technologies. Multimedia technologies consist of multiple modes, such as speech, images, writing, animations, and music (Kress 2010). When creating digital stories, some things must be considered: Input, output, time, difficulty, degree of participation and goals (Cole et al. 2000). Input means the instructional strategies which are used to facilitate the teaching-learning process. Output means how students can demonstrate their understanding; for example, in this study, they can imitate a digital storybook's sound. Teachers should consider the right time or length and difficulty grade of the story in the storybook. Digital storybooks should fit the students' knowledge properly. This consideration leads the students to participate actively during listening/reading the stories.

2.3 *Framework for promoting students' development*

There are some factors to promote students' development, such as (1) working in a community of learners; (2) using digital tools to make curriculum accessible; (3) and linking academic goals with real-world platforms (Price-Dennis et al. 2015). In the first point, to create community learners, students need to be active learners in their learning and share their knowledge, and take responsibility (Wells 2000). In the second point, as a previous study conducted by Price-Dennis et al. (2015), students had to share their thoughts and observe various modalities. The application and digital tools functioned as a mediator between individual and knowledge. In a previous study, they used Web 2.0 platforms as apps to create visual depictions. Meanwhile, in this study, the researchers used a story jumper as a digital storybook made initially by participants.

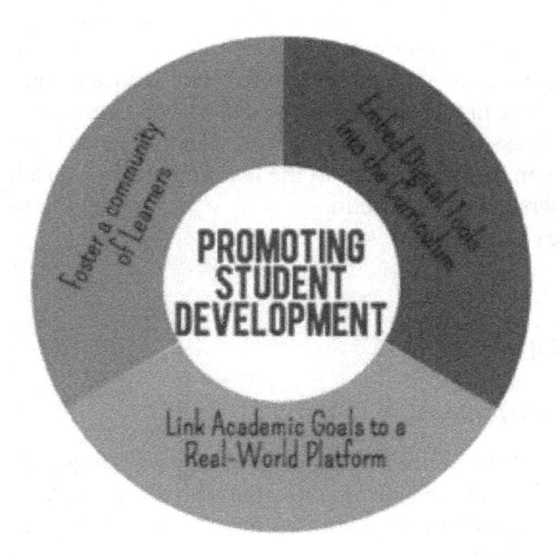

Figure 1. Framework for promoting students' development (Prince-Dennis et al. 2015).

3 RESEARCH METHOD

This research was exploring the practice on the digital storybook of special education needs students. In this research, the researcher used a qualitative research narrative inquiry. The narrative inquiry focused on the stories about self-lived experiences. These experiences consisted of past events, current events and future events (Barkhuizen et al. 2014). For several meetings, the researchers conducted a workshop with a total participation of sixty teachers. The researchers conducted purposive sampling to get more in-depth data from these participants by focusing on two participants. In this study, the selected participants were two teachers who implemented digital storybooks to teach English to their special education needs students.

The researchers used the interview to collect the data. The researchers used oral narrative data collection by conducting the semi-structured interview. To analyze the data, the researchers used thematic analysis of narrative inquiry. The thematic analysis compares narratives in the data set and highlights individual differences (Barkhuizen et al. 2014). The analysis of the interview used transcription, coding and analysis. The analysis was explored in detail the stories in terms of their content containing who, where and when which elaborated the context. All names in the data are pseudonyms.

4 FINDINGS AND DISCUSSION

As a result, the researchers chose a short story based on teachers' own experiences while implementing digital storybooks to their special education needs. The teacher made digital storybooks. These stories are explained with digital storybooks, technology and its implementation. These stories are collected from in-depth interviews between the researchers and the teachers.

4.1 *Participant 1 (P-1), a SEN teacher, 15 years of experience*

"I feel that technology creates different vibes during my teaching activity. Without technology, the teaching-learning activity becomes less interesting in a conventional way, and the teacher should make a greater effort to create good vibes and a positive atmosphere during online class. For example, when I taught using printed books, students focused on my gestures during offline class. Thus, students forgot the content of the story after I ended my gestures. However, using technology (digital media) creates a better atmosphere: the activity becomes more and more interesting. Besides that, animation and sounds on the digital storybook make the story more alive."

"To build engagement, I ask the students to imitate the voice from the storybook. This activity aims to make students more fluent and clearer because my students have difficulty speaking. Besides that, my students are getting more excited about their reading interests. This digital storybook is easy to use and they are happy to use it."

4.2 *Participant 2 (P-2), a SEN teacher, 11 years of experience*

"Rather than using conventional practice, my students prefer using technology, in this case, digital storybooks. The digital storybook has many modes that help my teaching activity. Voice, sound, and pictures are modes that exist in the digital storybook. These modes help me to build a positive atmosphere in my classroom. This positive atmosphere creates happiness in my class."

"To build the engagement, I try to maintain the interaction. I ask their opinions, which relate to the storyline. Besides that, I feel that a digital storybook helps my students to learn the materials easier. Since it has voices that my students can imitate, it creates happy vibes on our online teaching-learning activity. I believe that once we ask the students to practice, they will feel less bored. Voices or sounds on digital storybooks can also make my students express their feelings better."

5 DISCUSSION

"*... during offline class, when I taught using the printed book, students tended to focus on my gestures. Thus, students forgot the content of the story after I ended my gestures*" (P-1). Participant 1 stated that she only used a printed book for her class before she knew the digital storybook. During the teaching-learning process, she mostly used her gestures while reading some stories to students; however, she found that her students forgot the story's content because they focused on the teacher's gestures. "*... using technology (digital media, creates a better atmosphere, the activity becomes more and more interesting*" (P-1). Participant 2 (P-2) also stated the same point: "*Rather than using conventional practice, my students prefer using technology, in this case, the digital storybook*" (P-2).

Discussing students' interests, both participants mentioned modes on the digital storybook. Kress (2010) stated that digital storybooks support telling stories activities using multimedia technologies. Multimedia technologies consist of multiple modes such as speech, images, writing, animations and music. "*... animation and sounds on the digital storybook make the story more alive*" (P-1). To check her students' focus and understanding of the class, participant 1 kept the interaction and engagement to her students. This statement is also in line with Participant 2, who said, "*Rather than using conventional practice, my students prefer using technology, in this case, the digital storybook*" (P-2). "*I believe that once we ask the students to practice, they will feel less bored. Voices or sounds on digital storybooks can also make my students express their feelings better*" (P-2).

6 CONCLUSION

Special education needs students are like general students. They are also familiar with the technology. Moreover, they can also use technology in this pandemic, where all the teaching-learning process should be done at home. Creating teachers' storybooks that are originally taken from the curriculum helps them create creative and innovative materials to be learned by their students. Digital storybooks are easy to access. Modes on digital storybooks do help teachers and parents to deliver the content of the story. Sounds, music, moving images, color and other modes are part of multimodality to support exciting materials. Learners tend to have more interest when it comes to the technology used. However, a guide from an adult is still needed. The combination of technology and guidance from adults create good interaction. The interaction may be various, such as asking their opinion, asking to imitate, asking similarity, asking difference, or others.

REFERENCES

Barkhuizen, G., Benson, P., and Chik, A. (2014). Narrative inquiry in language teaching and learning research. New York: *Routledge*.

Biklen, D. (2000). Constructing inclusion: Lessons from critical, disability narratives. *International Journal of Inclusive Education, 4*(4), 337–353.

Bray, L., Mrachko, A., & Lemons, C. (2014). Standardized writing opportunities: A case study of writing instruction in inclusive classrooms. *Teachers College Record, 116*(6), 1–40.

Chong, A.P., & Shaffe, M.D. (2016). Relationship between Special Education (hearing impairment) teachers' Technological Pedagogical Content Knowledge (TPACK) and their attitudes toward ICT integration. *Seminar Antarabangsa Pendidikan Khas Rantau Asia Tenggara Siri Ke-6, 2016, February*, 313–319.

Cole, S., Horvath, B., Chapman, C., Deschenes, C., Ebeling, D. G., & Sprague, J. (2000). Adapting curriculum&instruction: A teacher's desk reference (2nd ed.) *Bloomington, IN: Indiana Institute on Disability and Community, p. 39.*

Jones, S., & Enriquez, G. (2009). Engaging the intellectual and the moral in critical literacy education: The four-year journeys of two teachers from teacher education to classroom practice. *Reading Research Quarterly, 44*(2), 145–168.

Mills, K. A. (2010). A Review of the "Digital Turn" in the new literacy studies. *Review of Educational Research, 80(2), 246–271.*

Oyler, C. (2011). Teacher preparation for inclusive and critical (special) education. *Teacher Education and Special Education: The Journal of the Teacher Education Division of the Council for Exceptional Children, 34*(3), 201–218.

Price-Dennis, D., Holmes, K. A., & Smith, E. (2015). Exploring digital literacy practices in an inclusive classroom. *Reading Teacher, 69*(2), 195–205.

Sleeter, C. E. (1986). Learning disabilities: The social construction of a special education category. *Exceptional Children, 53*(1), 46–54.

Standford, Pokey, CrCrowe, M. W., Flice, H., & Crowe, M. W. (2010). Differentiating with technology. *Teaching Exceptional Children Plus, 6(4), 1–9.owe et al., 2010)*

Williams, P. (2006). Exploring the challenges of developing digital literacy in the context of special educational needs communities. *Innovation in Teaching and Learning in Information and Computer Sciences, 5*(1), 1–16.

Wells, G. (2000). Dialogic inquiry in education: Building on the legacy of Vygotsky. In C.D. Lee & P. Smagorinsky (Eds.), Vygotskian perspectives on literacy research: Conducting meaning through collaborative inquiry (pp. 51– 85). New York, NY: *Cambridge University Press.*

Post Pandemic L2 Pedagogy – Adi Putra & Arifah Drajati (Eds)
ISBN 978-1-032-05807-8

The use of the flipped classroom approach to teaching English grammar

Nur Sehang Thamrin & Rofiqoh
Universitas Tadulako, Indonesia

ABSTRACT: Many previous studies prove the benefits of a flipped-classroom approach contributes to EFL students' learning achievement. This research aims at investigating the use of the flipped classroom approach in teaching grammar. The quasi-experimental research design was carried out on the first-year undergraduate students in Central Sulawesi. We used a purposive sampling technique to select a research sample, using two classes with 72 students with the lowest grammatical knowledge achievement. Thirty-six students represented the experimental group, in which we implemented the flipped classroom approach. In comparison, another 36 students of the same level represented the control group with the same teaching material, but without the flipped-classroom approach. The statistical analysis for the post-test reveals that the experimental group participants outperformed the control group. This finding strengthens the effectiveness of the flipped classroom approach as one of the language learning-teaching approaches.

Keywords: EFL, flipped classrooms, teaching grammar

1 INTRODUCTION

It could not be denied that grammar knowledge helps EFL learners to use English well in verbal or written form; otherwise, their language development will not succeed (Bao & Sun 2010; Widodo 2006). Learning grammar is challenging for most EFL students. It requires conceptual language knowledge and practices (Abdulmajeed & Hameed 2017; Liu et al. 2018; Thamrin et al. 2019;. Using ICT teaching approaches can solve this issue, such as a flipped-classroom approach. Chuang et al. (2018) claim that the flipped classroom approach is a form of blended learning model which facilitates online delivery of various learning resources. This model has given rise to a transformative instructional model where students learn course contents at home or outside the classroom in depth. (Albahuath 2020; Bergmann & Sams 2012; Floris 2014) The classroom is the place of discussing and practicing the concept of language with peers. Initiating to flip the grammar class encourages students–centered interaction and communicative grammar practices.

The implementation of the flipped classroom approach is not a new approach in EFL teaching-learning process. Several studies reported the flipped classroom approach's benefit on promoting students' motivation, autonomous learning, and critical thinking (Su Ping et al. 2019; Xiao et al. 2018; Zou & Xie 2019). Besides, Turan and Akdag (2019) had reviewed 43 articles regarding the implementation of the flipped classroom approach in English as SFL published in many reputable journals worldwide disclose that the study of this field increases significantly with various research methods. Still, most of them are on language skills, speaking, and writing. Despite the knowledge gained through the findings above, no study, to our knowledge, has explicitly examined the use of the flipped classroom approach in grammar class for EFL students in the field of research as the researchers believe that it helps teachers and students accomplish the instructional learning goals.

The research results on applying the flipped classroom approach in teaching-learning grammar lead to the enrichment and updating of English teaching practices regarding teaching-learning English grammar as a foreign language in Indonesia. Teachers provide materials in the form of videos and attractive slides shared in digital learning platforms, such as Youtube, Google classroom,

DOI 10.1201/9781003199267-10

or any form of learning management system that enable students to study anytime and anywhere. For example, it integrates the flipped classroom approach in the curriculum as teaching-learning strategies, especially in this research field. Therefore, this research investigates the flipped classroom approach in teaching grammar class on Passive Voice to EFL students at the tertiary level. There are several reasons why flipping the classroom is applied in EFL grammar lesson. Firstly, this paper intends to investigate if using the flipped classroom approach affects the students' learning outcomes; secondly, whether there is any difference in the students' learning outcomes between the group using the flipped classroom approach and the counterpart using the conventional learning approach.

2 LITERATURE REVIEW

Grammar is a set of structured rules governing the composition of sentences, phrases, and words in any language. It supports learners to use other language skills (Widodo 2006). However, English grammar has a different pattern than Indonesian grammar, which leads to a strong effort for teachers and students during the learning process. At the same time, the deficiency of practice times and interactive teaching materials are their barriers, which can basically be solved through ICT (Cårdenas-Claros & Oyanedel 2015). Learning grammar requires an understanding of abstract concepts and practices.

A flipped classroom approach is a model of blended learning that combines synchronous learning with asynchronous independent learning (Bergmann & Sams 2012; Cornelius & Gordon 2018). Using various instructional multimedia in the form of learning videos from YouTube, teacher-made video lectures, PowerPoint presentations, and online quizzes are regarded as an effective way to strengthen students' knowledge before in-class interaction. This situation contrasts with non-flipped models that students come into class with confused thoughts or even without any prior knowledge. The teachers must explain the material before asking them to practice the language used. Tang et al. (2017) examined the comparison between the flipped classroom and lecture-based classroom revealed that the flipped classroom approach promoted students' learning motivation, improved students' understanding of the course materials, and enhanced students' communication skills and clinical thinking. In an EFL context, Fathi and Rahimi (2020) conducted a study to investigate the flipped classroom's impact on learners' writing complexity, accuracy, and fluency. The findings support that the flipped classroom significantly developed students' writing skills and outperformed the non-flipped classroom learning achievement as it allowed for more significant differentiation of learning procedures.

The results of previous studies show the positive effect of the flipped classroom approach on student learning outcomes because it facilitates teachers to provide materials learned before the in-class meeting. Students are asked to review the material at home and even make notes that can be discussed in class. Whereas learning in the classroom is more effective in investigating whether students understand the material given and can apply it in written or spoken language. It can be concluded that learning with the flipped classroom approach is done in a way where synchronous learning occurs in real-time in the classroom. Learners interact with a teacher and classmates and receive feedback at the same time.

Meanwhile, asynchronous learning is learning that is more independent. Content is usually accessed through some form of media on digital platforms. Students can choose when they learn, and also, they can ask questions in the comments column and share their ideas or understanding of the material with teachers or classmates.

3 RESEARCH METHOD

The participants were 72 students with 44 females and 28 males aged 19 to 20 from the two parallel classes of six classes of the first-year English Department students taking Pre-Intermediate

Communicative Grammar Skills in one university in Palu, Central Sulawesi. The participants had studied English for at least six years through high school. The sample was recruited purposely by considering their low performance in grammar class. Class E and F had the lowest grammar mean scores of the previous grammar lesson among the six parallel classes.

This research aims at investigating whether the use of the flipped classroom approach develops the EFL students' grammar mastery. It employed a quantitative research approach, particularly quasi-experimental research design (Johnson & Christensen 2000). Thirty-six students acted as the experimental group who received the flipped classroom approach as the teaching-learning instruction. On the other hand, another thirty-six students represented the control group who were instructed by the same teaching material without the flipped classroom approach. The same lecturers taught both classes. The data collection instrument is a test, a pretest, and a posttest that included a list of grammatical competencies. The statistical method of paired-sample test and Mann-Whitney U test is used to analyze the data.

4 FINDINGS AND DISCUSSION

This part presents the answers to the stated research questions by implementing the statistical analysis. A paired-sample test was used to test the first research question regarding whether using the flipped classroom approach affects students' learning achievement. The Mann Whitney U test was run to explore to compare learning achievement between the experimental and control group.

First, there is a change in grammar performance, as measured by a timed grammar task, from the pre-test to the post-test for the experimental group after experiencing the flipped classroom. The analysis starts with presenting the descriptive statistics to explain whether there was a difference between the pretest and posttest of the experimental and control groups.

Table 1. Descriptive statistics of the experimental and control group.

	N	Minimum	Maximum	Mean	Std Deviation
Pretest Experiment Group	36	5	83	41.06	17.340
Post-Test Experimental Group	36	10	100	72.28	20.939
Pre-Test Control Group	36	0	95	41.81	29.207
Post-Test Control Group	36	25	100	61.53	26.317
Valid N	36				

Table 1 presents no significant difference between the experimental and control groups' mean scores before the treatment began. The mean score for the experimental group is higher than the control group, which means implementing the flipped classroom approach has positive effects on the students' grammatical competence. To prove whether the difference is significant or not, the researchers need to interpret the results of the paired sample t-test.

Table 2. T-test for experimental group.

			95% Confidence Interval of the Difference				
mean	Std. Deviation	Std.Error Mean	lower	upper	t	df	Si.(2-tailed)
−14.95333	8.96710	1.49452	−17.98736	−11.91930	−10.005	35	.000

Table 2 shows that Sig. 000 < p 0.05. It can be said that there is a relationship between the pretest and posttest variables, which means there is an effect of using the flipped classroom in improving student learning outcomes. The flipped classroom approach integrates synchronous and asynchronous learning conditions. It starts from the asynchronous phase, where students access and learn the same material through LMS and a discussion via WhatsApp at a different time. In contrast, the synchronous step allows students to attend a lecturing class at the same time. (Glesbers et al. 2014) state that "a combination of synchronous and asynchronous in e-learning would seem desirable to optimally support learner engagement and the quality of students learning (pg.30)." In the former phase, the teachers distribute the learning materials before the offline class in the form of slides with sound, links of learning videos from Youtube, and tasks which the students access at home on their own time and space. After learning the passive voice material through video, the students do the tasks about finding errors of passive voice sentences or transforming active voice to passive voice. Besides, the flipped classroom allows the students to pause and rewind the teacher, which they cannot do during face-to-face interaction (Bergmann & Sams 2012). It means as students may watch the videos and read the material as many as they want. The students may also discuss with the teachers via WhatsApp Group to strengthen their understanding of the offline class's topic. The flipped grammar classroom-based approach motivates the students to learn and promotes their autonomous learning (Han 2015). In the traditional class, the best students dominate the learning activities by raising their hands and speaking up as soon as they know how to do the task. It cannot be denied that the class consists of the mix-ability of the students. Some students can catch the lesson fast, while others struggle to understand the lesson better, so the struggling students show their disengagement during the lesson. However, after having the material learned at home, the struggling students show their active participation in the class. Based on our experience during the teaching journey, one of the struggling students with the initial name "MR" did not show motivation during the previous semester's learning process. Yet, when the flipped classroom was implemented in his class as the teaching instruction, he became active during the teaching journey.

In the second phase, face-to-face interaction in the classroom, the teaching-learning activities focused mainly on language use. Since the students have prior knowledge, they never hesitate to show their performance. Their active learning was seen throughout the teaching-learning process. Most of the teaching hours are devoted to the students' activities. It can be concluded that the flipped classroom approach also promotes student-centered learning (Tang et al. 2017).

Furthermore, to strengthen the result of the students' performance between that learning grammar through the flipped classroom approach and those with the traditional way, based on their scores in the pretest and posttest, the researchers applied Mann-Whitney Test.

Table 3 reveals that the comparison between the experimental and control groups' scores in the post-test was different even though it was too weak; 0,04 < the p 0.5. It can be concluded that there were differences in grammar mastery between students using the flipped classroom and those counterparts using conventional methods. The argument to explain this case is the domination of teacher-centered learning, which gives limited time for students to practice language use. If the teacher forces students to practice, the best students practise. As a result, the weak students became demotivated. The condition became worse as the students were not getting used to finding other sources themselves. They only relied on the reading sources given by the lecturer.

The present study gives pedagogical implications about using the flipped classroom approach in teaching English grammar to EFL students. Although both experimental and control groups have shown their achievement after learning passive voice for seven meetings on the post-test,

Table 3. Test statistics of Mann-Whitney Test.

	Students' Grammar Mastery
Mann-Whitney U	466.500
Wilcoxon W	1132.500
Z	2.058
Asymp Sig (2-tailed)	0.040

the measurement gains were high only for the experimental group whose teaching instruction was assisted by using the flipped classroom approach.

5 CONCLUSION

The principal objectives of this study are twofold – firstly, to investigate the effect of the flipped classroom approach in teaching grammar to EFL students, and secondly, to compare the learning achievement of EFL students in Sulawesi, Indonesia, who were exposed to the flipped classroom approach and the lecturing approach. This study's objectives were assessed based on the students' mean scores on the pretest and the post-test given to the experimental and control groups. The result reveals that the flipped classroom approach helps EFL students improve their learning achievement. Furthermore, this approach encourages the students to learn independently, as they can access the material on their own space and time through their smartphone, tablet, or laptop. They may discuss with the lecturers as soon as they find something unclear about the material provided to them.

While this study supports the benefits of using the flipped classroom approach in teaching English grammar in Indonesia, this study has a limitation in generalizing other research scopes. It involved only 76 students as participants. The number of samples was too small as a basis for generalizing the result. For further research, the integration of qualitative survey methods can be conducted to investigate EFL students' perspectives towards the use of the flipped classroom approach in practicing language use. Regarding the number of research participants, for further study, the participants should be a lot and taken from several institutions to get various participants.

REFERENCES

Abdulmajeed, R.K., & Hameed, S.K. (2017). Using a linguistic theory of humor in teaching English grammar. *English Language Teaching,*107(2), 40–47.

Bao, J., & Sun, J. (2010). English grammatical problems of Chinese undergraduate students. *English Language Teaching*, 3(2), 48–53.

Bergmann, J., & Sams, A. (2012). Flip your classroom; reach every student in every class every day (1st ed.). USA: Courtney Burkholder.

Càrdenas-Claros, M., & Oyanedel, M. (2015). Teachers' implicit theories and use of ICTs in the language classroom. *Technology, Pedagogy, and Education*, 25(2), 207–225.

Chuang, H. H., Weng, C. Y., & Chen, C. H. (2018). Which students benefit most from a flipped-classroom approach to language learning? *British Journal of Educational Technology,* 49(1), 56–68.

Cornelius, S., & Gordon, C. (2018) Providing a flexible, learner-centered program: Challenges for educators. *Internet and Higher Education,* 11(1), 33–41.

Fathi, J., & Rahimi,M.,(2020).Examining the impact of flipped on writing complexity, accuracy, and fluency: a case of EFL students. *Computer Assisted Language Learning.Latest Articles,* 1–40.

Floris, F. D. (2014). Using Information and Communication Technology (ICT) to enhance language teaching & learning: An interview with Dr. a. Gumawang Jati. *TEFLIN Journal,* 25(2), 139–146.

Glesbers, B.; Rienties, B.; Tempelaar, D; & Glijselaars, W. (2014). A dynamic analysis of the interplay between asynchronous and synchronous communication in online learning: The impact of motivation. *Journal of Computer Assisted Learning*, 30(1), 30–50.

Han, Y. J. (2015). Successfully flipping the ESL classroom for learner autonomy. *NYS TESOL Journal*, 2(1), 98–109.

Johnson, B., & Christensen, L. (2000). Educational Research: Quantitative and Qualitative Approaches.Needham Heights, Massachusetts: A Pearson Education Company.

Liu, C., Sands-Meyer, S., & Audran, J. (2018). The effectiveness of the students' response system (SRS) in English grammar teaching in a flipped English as a foreign language (EFL) class, 27(8), 1178–1191.

Su Ping, R.L., Verezub, E., Badiozaman, I.F.A., & Su Chen, W. (2019). Tracing EFL students' flipped classroom journey in a writing class: Lessons from Malaysia, 57(3), 305–316.

Tang, F. et al. (2017). Comparison between the flipped classroom and lecture-based classroom in ophthalmology clerkship. M*edical Education Online*, 22(1), 1–9.

Thamrin, N.S., Suriaman, A., & Maghfirah (2019). Students' perception of the implementation of Moodle web-based in learning grammar. *IJOTL*, 4(1), 1–10.

Turan, Z., & Akdag-Cimen, B. (2019). Flipped classroom in English language teaching: a systematic review. *Computer Assisted Language Learning,* 33(5–6), 590–606.

Widodo, H. P. (2006). Approaches and procedures for teaching grammar. E*nglish Teaching: Practice and Critique*, 5(1), 122–141.

Xiao, L., Larkins, R., & Meng, L. (2018). Track effect: Unraveling the enhancement of college students' autonomous learning by using a flipped-classroom approach. *Innovations in Education and Teaching International*, 55(5), 521–532.

Zou, D., & Xie, H. (2019). Flipping an English writing class with technology-enhanced just-in-time teaching and peer instruction. *Interactive Learning Environment,* 27(8), 1127–1142.

Post Pandemic L2 Pedagogy – Adi Putra & Arifah Drajati (Eds)
 ISBN 978-1-032-05807-8

Teacher – parent partnerships in English virtual learning

Farikah, Moch Malik Alfirdaus, Hari Wahyono, Mursia Ekawati & Dwi Winarsih
Universitas Tidar, Magelang, Indonesia

ABSTRACT: This case study approach investigated the contribution of parent-teacher partnerships in English Virtual learning during the COVID-19 period. The parent-teacher engagement was seen as a critical factor in children's educational success especially during the COVID-19 period. The case study was applied in conducting this study. Virtual observation, questionnaires, and teacher's experiences were used as sources of information in collecting the data. This study was administered to five parents of elementary school children in Indonesia aged 7–12 years old and two English teachers of elementary school. The data were also coded based on Reissman (2008). The finding had shown that parent involvement in English class gave benefits not only to the child but also the parents and teacher. It provided two-way information direction from the teacher to the parents about the child's classroom achievements and personal and from the parent to the teacher dealing with the complementary elements in the home environment.

Keywords: Case study, English virtual learning,teacher-parent partnerships.

1 INTRODUCTION

The COVID-19 Pandemic outbreak has made the world society busy including Indonesia. Almost all aspects of life have profoundly changed including education in Indonesia. To anticipate the outbreak from spreading more widely, the Indonesian government has made some regulations. Social distancing and physical distancing regulation are created due to the increasing number of people infected with COVID-19. These are followed by the school from home (SFH) policy for the students.

In line with the SFH policy for the students, some learning systems have been created. It has forced the school to implement distance education or online learning. On the other hand, the implementation of SFH policy has made some obstacles. Some challenges are faced including instructional problems. The problems faced by the students motivate the teachers, and parents to take big parts in this context. Their contribution becomes the key that cannot be neglected in supporting the students' success during the COVID-19 Pandemic outbreak era. Some studies reported that parents' involvement gives some benefits in education. As is mentioned by Eldridg (2001), in education, parent involvement gives benefits not only to the child but also to the parents and teachers In addition, many reported that teacher-parent partnership gives a high contribution to the students' learning success. They play a big role in education (Murray et al. (2013), Epstein (2018), Loughran (2008), Gisewhite et al. 2019)). The value of collaboration between parents and educators is well-recognized. The parent-teacher engagement was seen as a critical factor in children's educational success especially during the COVID-19 period. This article will explore the contribution of parent-teacher partnerships in English Virtual learning during the COVID-19 period. What roles are played by the parents and It also describes the parent-teacher partnerships in English virtual learning during the COVID-19 period.

 DOI 10.1201/9781003199267-11

2 LITERATURE REVIEW

In the view of partnership, some previous studies confirmed that in education, the partnership between teachers and parents is needed. This partnership can facilitate student success effectively (Gonzalez-DeHas et al. 2005; Simpkins et al. 2006). Thomas et al. (2019) confirmed that parents play an important role in students' self-regulated learning development. The Parent-involvement model gives benefits greatly in succeeding in the school program (Cooper & Maloof 1999). The need for this relationship becomes essential nowadays because of a pandemic situation. The minister of education and culture of Indonesia through circular no. 4 year 2020 announced the implementation of education policy in a state of emergency because of the coronavirus disease (COVID-19). The minister mentioned that the learning processes at home are applied with the following points.

1. The learning process from home is held to give a meaningful learning experience to the students.
2. The focus of learning from home is such as the skill to face the pandemic situation.
3. The activities at home can be varied based on the students' condition and interest.
4. The students' reports or the product of learning from home can be evaluated qualitatively and given feedback by the teachers directly.

In line with the above policy, the use of online learning in the teaching-learning process of English in Elementary school is one of the wise solutions. Otherwise, this cannot guarantee that all the processes can run well. The availability of the facilities becomes one of the important keys to the success of online learning at home. Besides, in online learning during school from home policy, the parents' support cannot be neglected. The communication between parents and teachers in educational settings plays a critical role in building and supporting healthy interpersonal relationships (Gisewhite et al. 2019). Building positive relationships with parents becomes one of the most important aspects of teaching. It is a possible strategy for improving the classroom management environment. Effective parent-teacher communication is essential for a teacher to be successful. In addition, it is also essential for a student to make progress. To educate youth effectively in schools, families, and communities must become a full partnership in the process or it is called School, family, and community partnerships. (Hoover-Dempsey et al. 2001). Most teachers think about having a good relationship with parents. This view is in line with Epstein et al. (2002) who described parents' and teachers' cooperation which consisted of six types (forms) of partnership. They are Parenting, Communicating, Volunteering, Learning at Home, Decision Making, and Collaborating with the Community.

3 METHOD

The case study involved five parents of elementary school children in Indonesia aged 7–12 years old. The five parents participated voluntarily. In addition, they have various socioeconomic backgrounds. Besides, two English teachers were also invited to participate in this study to give their experiences while supervising their children and also their students in learning English. This study was conducted for one semester.

The purpose of this case study was to describe the contribution of parent-teacher partnerships in English Virtual learning during the COVID-19 period. The case study was implemented since the writer wanted to describe in detail what the parents did and what roles they played in their children's teaching-learning process during virtual learning because of the COVID-19 Pandemic outbreak. In addition to that, this study also explored the parents' experience when they supervised their children in English virtual learning. Virtual observation, questionnaires, and teacher's experiences were used as sources in collecting the data. The study involved three phases. Firstly, five parents of elementary school children as the participants of the study were given consents form through a Google form. After that, they were given some additional information about the study (i.e. purposes of the study), and finally, the virtual interview and virtual questionnaires began. The questionnaires and interviews took place at the end of the first semester of virtual learning. Following the procedures of 'narrative interviewing' (Reissman 2008), the virtual interviews were conducted in a casual manner. The interviews were started by asking some questions about her general experience during the teaching-learning process at elementary school during COVID-19 time. This was followed by some specific questions requesting them to elaborate on their experience and perception in joining an English virtual learning class. The stories were also coded based on Reissman (2008).

4 FINDING AND DISCUSSION

In line with the problems, the analysis of the data produced four themes. Those are the parents' help in assisting the students (children) in virtual learning, encouraging the students (children) to be autonomous learners in virtual learning, helping fulfill the students' need in virtual learning and the last is dealing with the students' experience in virtual learning.

4.1 *Assisting the students (children) in virtual learning*

The first aspect of the teachers-Parents' Partnership in virtual learning is assisting the students (Children) in virtual learning. Based on the questionnaire data, most of the parents actively helped their children in virtual learning. The facts were supported by questionnaire data which mentioned that 65.63% of the students were assisted by their parents in virtual learning. In addition, 34.37% of them stated that they did their learning by themselves. They were not assisted by their parents or families. Further, based on interview data, it could be inferred that more than 50% of them could not do virtual learning by themselves. They needed assistance.

In addition to the questionnaire data, based on the parents' interview data, they helped their children in virtual learning because of internet connection, time-consuming in doing assignments, understanding the technological tools, and understanding of the materials. Below are the excerpts of the parents' interview

Participant a: *I find difficulties because the learning process cannot be done directly with the teacher and the internet connection is poor.*

Participant b: *I find difficulties in listening and understanding the teacher's voice because she delivers in English so I don't understand.*

Participant c: *I don't understand because no face to face interaction so it's difficult for me to understand the pronunciation.*

4.2 *Encouraging the students (children) to be autonomous learner in virtual learning*

Virtual learning forces the students and parents to learn. It was based on the research data. 40.63% of the students stated that they studied by themselves during virtual learning. 31.25% were always helped by their parents. And 28.12% of the students were sometimes helped by their parents and sometimes did by themselves in their learning. To support the questionnaire data, below are the excerpts of the Parents' Interview.

Participant a: *Yes, my son does online assignment himself because we as his parents must work every day.*

Participant b: *My daughter usually does the online assignment herself otherwise she will ask me if she doesn't understand certain materials.*

Participant c: *Of course, my daughter always does her homework or school assignment with me (mother) because in my house, I have only one handphone and I bring it myself.*

Helping the students' need in virtual learning

Parents' roles cannot be neglected in English virtual learning. They contributed highly towards the success of their children's learning. Based on the students' questionnaires, 6.25% of the students stated that their parents help them in arranging and pronouncing the English words, 37.5% guiding (teaching) them, and 56.25% helping them in translating the words.

The excerpts of the parents' interview below are used to support the questionnaire data.

Participant a: *If there are difficult words that can't be understood by my daughter, we as her parents will always help her consulting to the dictionary or google translate*

Participant b: *When accompanying my son learning, sometimes I don't understand the vocabularies so I ask him to consult the dictionary or the internet.*

Participant d: *I just give her motivation to learn since I don't understand the materials.*

4.3 *Experiencing virtual learning*

Virtual learning was a new experience for almost all of the elementary school students in Indonesia. Some problems they found are the following. 31.26% of the students found signal as the crucial problem. 25% of the students found problems on comprehension problems. 3.13% found technical or problems dealing with the lack of mastering technology. 3.13 % found problems dealing with the availability of the facilities.

Besides some problems, some of the students stated that virtual learning is communicative and interesting model of learning. 21.88% stated that virtual learning is an interesting model of learning. 15.63% stated it as a communicative activity of learning. Below are the excerpts of the parents' interview

Participant a: *I am very very happy when I find the questions are easy, friendly signal, I get trouble when I don't find a good signal, no data voucher or her father brings the handphone. We only have one Hp.*

Participant b: *It's difficult when the tasks are difficult because there no face to face interaction with the teacher, but the positive side is, more time to have interaction with my child because of learning at home.*

Participant c: *Learning at home makes us clever because we are forced to learn also. It's challenging but interesting.*

Apart from that, based on teachers' points of view, virtual learning challenged the students and their parents. Below are teachers' expressions during virtual learning class.

Participant a: *As a teacher, I handle my teaching-learning activities by zoom application, besides I prepare students' worksheets to help my students learning at home. I deliver the information through the parent's WhatsApp group. I also encourage parents to take advantage of cellphones for positive activities such as virtual learning.*

Participant b: *As a first-grade teacher, the model of online learning cannot be implemented because of natural conditions, no signals, and some of the students' parents have no cellphone. So from that, we do the tour of the students' house. I make the students into some groups. In learning English, I focus on pronunciation. Besides, I use teaching aids in the forms of letters and numbers. I begin by teaching them spelling and reading slowly.*

From the above interviews and questionnaire results, it can be concluded that in virtual learning there must be a partnership between teachers and parents. This kind of partnership can provide two-way information direction. It is from the teacher to the parents about the child's classroom achievements and personal and from the parent to the teacher about the complementary elements in the home environment. This is in line with Fehrmann et al. (1987) who reported that parents' involvement gives direct effects on the students' grades through homework and TV times. The parents can directly monitor their children's activities by monitoring what they are doing at school.

Based on the findings, it can be inferred that partnership between parents and teachers in virtual learning of their students gives a high contribution to the students' success in learning. Their cooperation makes a crucial role. Six types (forms) of a partnership are realized during virtual learning. They are Parenting, Communicating, Volunteering, Learning at Home, Decision Making, and Collaborating with the Community. This finding is also parallel with Swick's study (2003) which indicates the benefit of school, society, and family partnership in the students' learning. In addition, the above finding is also in line with Richardson and Newby (2006). It indicates that as students gain some experience with online learning, they can take more responsibilities for their own learning.

5 CONCLUSION

The finding had shown that parent involvement in English class gave benefits not only to the child but also the parents and teacher. The parents help in assisting the students (children) in virtual learning, encouraging the students (children) to be autonomous learners in virtual learning, and helping to fulfill the students' needs in virtual learning. It provided two-way information flow from the teacher to the parents about the child's classroom achievements and personal and from the parent to the teacher about the complementary elements in the home environment. In line with the finding, it is expected that the parents need to monitor and supervise their children's learning especially during COVID-19.

REFERENCES

Cooper, T. C., & Maloof, V. M. (1999). Parent involvement in teaching elementary-level Chinese, Japanese, and Korean. *The Journal of Educational Research, 92*(3), 176–183.

Eldridge, D. (2001). Parent Involvement: It's Worth the Effort. *Young children, 56*(4), 65–69.

Epstein, J.L. (2018). School, family, and community partnerships in teacher professional work. *Journal of Education for Teaching.*

Epstein, J. L., Sanders, M. G., Simon, B. S., Salinas, K. C., Jansorn, N. R., & Van Voorhis, F. L. (2002). *School, family, and community partnerships: Your handbook for action* (2nd ed.). Thousand Oaks, CA: Corwin.

Fehrmann, P. G., Keith, T. Z., & Reimers, T. M. (1987). Home influence on school learning: Direct and indirect effects of parental involvement on high school grades. *The Journal of Educational Research, 80*(6), 330–337.

Gisewhite, R. A., Jeanfreau, M. M., & Holden, C. L. 2019. A call for ecologically-based teacher-parent communication skills training in pre-service teacher education programs. *Educational Review*, 1–20.

Gonzalez-DeHass, A.R, Willems, P.P, & Doan Holbein, M.F. (2005). Examining the relationship between parental involvement and student motivation. *Educational Psychology Review*, 17(2), 99–123.

Hoover-Dempsey, K. V., Battiato, A. C., Walker, J. M., Reed, R. P., DeJong, J. M., & Jones, K. P. (2001). Parental involvement in homework. *Educational psychologist, 36*(3), 195–209.

Loughran, S.B. (2008). The Importance of Teacher/Parent Partnerships: Preparing Pre-Service and In-Service Teachers. *Journal of College Teaching & Learning*, 5 (8).

Menteri Pendidikan dan Kebudayaan Republik Indonesia Surat Edaran Nomor 4 Tahun 2020 Tentang Pelaksanaan Kebijakan Pendidikan Dalam Masa Darurat Penyebaran *Coronavirus Disease* (COVID-19).

Murray, M.M, Mereoju, M., and Handyside, L.M. 2013. Building Bridges in Teacher Education: Creating Partnerships with Parents. *The Teacher Education,* 48: 218–233.

Reissman, C. K. (2008). Narrative Methods for the Human Sciences. *Thousand Oaks, CA: Sage*.

Richardson, J.C., & Newby, T. 2006. The Role of Students' Cognitive Engagement in Online Learning. *The American Journal of Distance Education*, 20 (1), 23–37.

Sanders, M. G., & Epstein, J. L. (2005). School-family-community partnerships and educational change: International perspectives. *Extending educational change* (pp. 202–222).

Simpkins, S.D., Weiss, H.B., McCartney, K., Kreider, H.M., & Dearing, E. (2006). Mother-child relationship as a moderator of the relation between family educational involvement and child achievement. *Parenting Science & Practice, 6(1), 49–57.*

Swick, K. J. (2003). Communication concepts for strengthening family-school-community partnerships. *Early Childhood Education Journal, 30*(4), 275–80.

Thomas, V., Muls, J., De Backer, F., & Lombaerts, K. (2019). Exploring self-regulated learning during middle school: views of parents and students on parents' educational support at home. *Journal of Family Studies*, 1–19.

Post Pandemic L2 Pedagogy – Adi Putra & Arifah Drajati (Eds)
© 2021 Taylor & Francis Group, London, ISBN 978-1-032-05807-8

"Thank you, Teacher!": A critical reflective narrative of a foreign EFL teacher's career journey in Saudi Arabia

Kristian Adi Putra & Fahd Shehail Alalwi
Prince Sattam Bin Abdulaziz University, Al Kharj, Saudi Arabia

ABSTRACT: This study describes a critical reflective narrative of a Senegalese EFL teacher, Tafsir (a pseudonym), on his twenty-year teaching career in Saudi Arabia. We focused on Tafsir's reflections on his past, present, and future journey as a professional educator and what he thought to be the essential qualities a good EFL teacher needs to have. We collected the data through a series of in-depth semi-structured interviews and follow-up informal conversations. The collected data were analyzed using critical discourse analysis and framed with Vygotsky's sociocultural theory. In addition to highlighting the importance of passion for the profession and patience with some possible cultural and linguistic differences, Tafsir also emphasized the need for EFL teachers in Saudi Arabia to continuously learn and develop their pedagogical, professional, social, and personal competencies. These findings have implications for pedagogical practices and future studies, which we will discuss further in the article.

Keywords: Critical reflections, EFL, narrative inquiry, sociocultural theory, language teacher identity

1 INTRODUCTION

> While I was driving, a car behind me persistently tried to ask me to stop. Then I stopped. The driver got out of his car and called me "Mr. Tafsir" loudly. I was quite surprised and got out of my car too. He approached me and then hugged me tightly for five minutes or so. He said, "Thank you, Teacher! I just got my B.A. in English. It's all because of you." Everyone looked at us, as it was at noon and on a busy street during peak hour. (Tafsir, Interview, December 1, 2020)

Tafsir (a pseudonym) was a Senegalese EFL teacher who taught college students in different preparatory year programs in Saudi Arabia for twenty years. He shared the story above when we asked him about one moment in his teaching career that he thought to be a life-changing experience. He mentioned that one student failed in his class, who happened to be the son of one of the top administrators in the community college where he was working. In other courses, the student passed and got an A+. Later, the father asked Tafsir to give him an explanation about his son's grade. Tafsir provided all the student's performance data, including his attendance record, to justify why the student did not pass. Tafsir thought that the father would be furious. It turned out that he was wrong. The father thanked him and appreciated his honesty, integrity, and professionalism. He wanted Tafsir to teach his son again in the following semester. He also promised to monitor his son's progress of learning at home. In the end, the student changed and showed a significant improvement. He was also able to transfer to one of the best public universities in Saudi Arabia. Tafsir added how this experience always made him grateful for his career choice. He also used it as a reminder of the values that he needed to continuously hold as a professional educator, which was also something that he always wanted to share with his colleagues.

In line with Tafsir's story, in the last ten years, scholars have also given particular attention to the connection of the study of critical reflections (See Farrell 2019; Gun 2010) to the development of L2 teacher identity (See Barkhuizen 2016; De Costa & Norton 2017; Kayi-Aydar 2019). Teacher

DOI 10.1201/9781003199267-12

identity or "... who teachers are and what sort of experiences they bring to the classroom setting" (Higgins & Ponte 2017, p. 16) is the most crucial factor that ensures the classroom instruction quality and students' learning. This perspective comes from the argument that a complex L2 classroom dynamic is involved. Thus, it is not as simple as using a particular teaching methodology to improve teaching and learning and students' L2 competence and performance. Teachers' experiences and their ability to understand such a classroom's complexity and reflect critically on their daily practice of teaching before, while, and after the classroom sessions (Cirocki & Widodo 2019) are, in fact, much more pivotal.

That is to say, teacher identity continues to (re)develop throughout the teaching profession. Brookfield (2017) mentioned that such a (re)development process is shaped by continuous teachers' critical reflective practices, from students' feedback, colleagues' feedback, theories, previous studies, and personal experiences. For example, many previous studies have explored how pre-service teachers constructed and embraced their new identity as teachers while being in a teacher education program (Goktepe & Kunt 2020; Ruohotie-Lyhty & Pitkänen-Huhta 2020; Yuan et al. 2019). Other researchers also looked at how in-service, particularly novice, teachers continuously tried to reconstruct and renegotiate their beliefs and daily teaching practices (Farrel, 2008; Huang et al., 2019; Yuan, 2019). Some studies illustrated their teacher identity development and transition from an imagined teaching community (See Lave & Wenger, 1991) to a real teaching community of practice (Hong et al. 2017; Goktepe & Kunt 2020; Jiang et al. 2020). In general, all these studies provide insights into the dynamics of teachers' daily classroom routines and the academic and non-academic support that the teacher educators and school administrators could provide.

Sang (2020), in his review on language teacher identity studies, mentioned that further studies in this topic need to,

> "... examine how pre-service L2 teachers develop language teacher identity in professional socialization, and how language socialization influences L2 teachers' language teacher identity formation" (p. 6).

Sang's recommendation follows sociocultural perspectives (Vygotsky 1978). Human, in this case, teacher, learning takes place from their physical and social interaction with other people in their community of practice. It requires the internalization of skills, knowledge, attitudes, and values from such socialization. Throughout the journey of their teaching profession, teachers are, thus, expected to continuously (re)construct and (re)negotiate their professional identity through "the interpretation and reinterpretation of their life experiences, especially their language learning and teaching experiences" (Li 2020, p. 5).

While previous studies have commonly looked at the experiences of pre-service and novice teachers, in this study, we tried to describe the story of Tafsir (a pseudonym), an experienced non-native English speaking teacher (NNEST) in Saudi Arabia from Senegal. More specifically, this study aims to describe Tafsir's critical reflections on his experience teaching English in Saudi Arabia. We also detail his recommendations for EFL teachers who have just started teaching or are planning to pursue professional teaching careers in Saudi Arabia or other Middle Eastern countries. Learning from an experienced teacher's experiences, we expect to provide new insights on factors that make him survive as an ELT professional and challenges that pre-service and novice teachers might encounter and how to overcome them.

2 RESEARCH METHODOLOGY

This study's participant is Tafsir (a pseudonym), an experienced EFL teacher in Saudi Arabia from Senegal. At the time of the study, he was 62 years old and had taught EFL for first-year college students in Saudi Arabia for twenty years. The college classes in Saudi Arabia were separated based on gender, including in his current institution in the suburb of Riyadh. Therefore, for twenty years, he also worked with male colleagues and taught only male students.

This study uses narrative inquiry as to the research design. As explained by Barkhuizen et al. (2014), "Narrative inquiry can help us to understand how language teachers and learners organize their experiences and identities and represent them to themselves and others" (p. 5). We collected data through a series of semi-structured interviews conducted virtually via Zoom and follow-up informal conversations. The first two interviews focused on Tafsir's story of why he chose to teach as his profession and why he worked as an English language teacher in Saudi Arabia. In the following two interviews, we explored his teaching journey in Saudi Arabia and what made him last for twenty years. After we finished writing the initial manuscript, we allowed him to reflect on the story and provide feedback on what he wanted to add or delete. All the interview data were recorded, transcribed verbatim, coded, and analyzed using critical discourse analysis (Catalano & Waugh 2020; Fairclough 2013).

3 FINDINGS AND DISCUSSION

In what follows, we would like to describe two significant discourses that appeared in our data analysis results from the story of Tafsir's English language teaching career in Saudi Arabia. We focus on what he thought a professional English language teacher should look like and how to survive in the ELT profession in Saudi Arabia and other Middle Eastern countries.

3.1 *Tafsir's views on an ideal English language teacher*

> A good teacher should have a passion for teaching. They should love what they do. Teaching is our dedication to help students ready to face the reality of life. (Tafsir, Interview, December 5, 2020)

During the interview, Tafsir repeatedly emphasized the importance of teachers being passionate about teaching. It includes the teaching of knowledge, skills, attitudes, and values. For him, teaching is not just about making the students pass the test and get a good grade, but it is also about preparing the students to be back to society and face the real world. Therefore, teachers need to make sure that students graduate with quality academic aspects and good character traits.

Tafsir mentioned that teaching and learning goals should always be about equipping students with a lifelong and independent learner's knowledge and skills in such an ever-changing world. To teach the knowledge and skills well, teachers need to have professional and pedagogical competencies. Professionally, teachers should know what they are teaching. In the case of an EFL teacher, it is the knowledge about the English language and the skills to use the language. Pedagogically, they should also know how to deliver the lessons to their students effectively. They should engage their students in classroom activities and make their students excited and curious to learn more.

To teach attitudes and values, Tafsir added that teachers also need to have social and personal competencies. Teachers need to communicate well with their students, play a role as their students' supporters of learning, and be role models for their students. Tafsir mentioned that in his classrooms, he often shared humor with his students. He wanted his students to be comfortable with him and enjoy learning in his classrooms. In his opinion, a good teacher can create a good relationship with his students and a supportive classroom environment. At the same time, he also needs to nurture students' good character traits. While we shared about plagiarism in the final test of the writing course, Tafsir tried to elaborate it with similar stories that he had encountered in the past. He stressed the need for teachers to teach their students what was right and wrong and explain the reasons. To do that, students would know the consequences of their actions and try to avoid the same mistakes in the future.

3.2 *Tafsir's recommendation for new ELT teachers in Saudi Arabia*

Continuously learning and developing professionalism

In our four interviews with Tafsir, he continuously highlighted the importance of English language teachers in keeping learning and developing professionalism. For instance, during the second interview, he referred to how all the teachers in Saudi Arabia suddenly needed to switch to emergency remote teaching in mid-March 2020. He mentioned that during the transition, teachers needed to learn fast how to do everything online, starting from the tools, the teaching and testing materials development, the creative and engaging lesson delivery, and the administration of testing. Therefore, teachers needed to continuously look for professional development opportunities either facilitated by their institution or organized by external professional organizations and campuses.

Tafsir also gave another example of how teachers would always face different challenges and learning opportunities in their daily classroom practices and interactions with students, colleagues, and administrators. In one of the interviews, for instance, Tafsir said,

> In one semester, we might teach the same subject in two different classes. But they will never be the same. We have students with different [cultural and linguistic] backgrounds, personalities, and abilities. So the dynamics in the classrooms will also be different. That will allow you to think about various issues and challenges and strategies to solve them, making you continue learning and becoming a better teacher. (Tafsir, Interview, December 1, 2020)

Tafsir's statement above is in line with Farrell's study (2008). Farrel showed how eighteen English language teachers in Singapore wrote their critical reflections on critical incidents in their classrooms. The teachers also mentioned how such practice gave them opportunities to understand better their classrooms' issues and different strategies that they could try to solve.

Tafsir also pointed out that the debate about whether native English speaking teachers (NEST) or non-native English speaking teachers (NNEST) was the better teacher was no longer relevant in the ELT profession. What was more important, in his opinion, was the competencies, qualifications, and professionalism of the teachers. Although NEST currently still has some better privileges than NNEST in Saudi Arabia, in the future, Tafsir thought that eventually it would change and the three qualities above would be the one that matters. This perspective is in line with a lot of scholars, such as Galloway (2013; 2017) and Prabjandee (2019).

3.3 *Being resilient and able to adapt to a new social environment and academic expectation*

> I know people here have different personalities. Some people tend to be much more introverted. Some people can easily make friends and talk to new people. However, to survive, you need to be able to adapt and socialize well. You need to have the ability to ask and listen a lot while you are adjusting to a new life in a new place. (Tafsir, Interview, December 1, 2020)

Responding to what new expatriate teachers should anticipate when starting or planning to pursue a teaching career in Saudi Arabia or other Middle Eastern countries, Tafsir stressed the importance of being resilient and adapting to a new social environment and academic expectation. The social adaptation here includes the ability to get along well with people in the new place on and off-campus. Expatriate teachers in Saudi Arabia might experience a few cultural differences in addition to the language barriers. It will happen on campus when they are interacting with students, colleagues, and administrators and off-campus with taxi drivers, cashiers in grocery stores, apartment landlords, and some other real-life settings. Tafsir mentioned that one of the issues that many new faculty members faced on campus, especially those who did not speak and write in Arabic, was when they needed to complete some administrative paperwork and communicate some issues they encountered with some college administrators. While all colleagues spoke in English fluently, many administrators just spoke in Arabic. Therefore, sharing such problems with colleagues on campus who speak Arabic to ask for help will help. Survival Arabic skills will also be useful.

A new expatriate EFL teacher in Saudi Arabia also needs to be able to learn fast. They need to know the new system, expectations, and strategies to achieve them. Tafsir added that it could be very different from what they had experienced before. For example, while in other countries, EFL college instructors had the freedom to create their own teaching and testing materials, in Saudi Arabia, all of them were commonly standardized. He noticed that some instructors might have a tendency to disagree with such a policy. However, as a newcomer, Tafsir reminded that they need to adjust to it slowly and understand the reasons behind each of the policy. Once they settle, they can communicate it with policymakers and try to make it better. Such negotiation, in his opinion, will help them adjust to their new academic environment better.

4 CONCLUSION

This study described a professional NNEST twenty-years teaching experience in various preparatory year programs in Saudi Arabia. We highlighted his perspectives on the ideal EFL teacher's qualities and what an EFL teacher should prepare and continuously learn if they want to pursue a teaching career in Saudi Arabia or other countries in the region. In general, Tafsir highlighted the need for a professional English language teacher to have professional, pedagogical, personal, and social competencies. An EFL teacher also needs to continuously learn, develop their professionalism, and be resilient and able to adapt to a new working and social environment. This study provides insights for pre-service, novice, and in-service EFL teachers about what they need to prepare and continuously improve. It also gives college administrators a description of what, why, and how to support their teacher's professional development. Given relatively few participants in this study, further studies need to be conducted with many participants from different nationalities and working in some foreign countries. Such an analysis is to compare the experiences and insights about this issue from different perspectives.

REFERENCES

Barkhuizen, G. (Ed.). (2016). *Reflections on language identity research*. New York, NY: Routledge.

Barkhuizen, G., Benson, P., & Chik, A. (2014). *Narrative inquiry in language teaching and learning research*. New York, NY: Routledge.

Brookfield, S.D. (2017). *Becoming a critically reflective teacher.* San Francisco, CA: Jossey Bass.

Catalano, T., & Waugh, L. R. (2020). *Critical discourse analysis, critical discourse studies, and beyond.* Switzerland: Springer.

Cirocki, A., & Widodo, H. P. (2019). Reflective practice in English language teaching in Indonesia: Shared practices from two teacher educators. *Iranian Journal of Language Teaching Research, 7*(3), 15–35.

De Costa, P. I. & Norton, B. (2017). Introduction: Identity, transdisciplinarity, and the good language teacher. *The Modern Language Journal*, 101(S1), 3–14).

Fairclough, N. (2013). *Critical discourse analysis: The critical study of language*. New York: NY: Routledge.

Farrell, T. S. (2008). Critical incidents in ELT initial teacher training. *ELT Journal, 62*(1), 3–10.

Farrell, T. S. (2019). Standing on the shoulders of giants: Interpreting reflective practice in TESOL. *Iranian Journal of Language Teaching Research, 7*(3), 1–14.

Goktepe, F. T., & Kunt, N. (2020). "I'll do it in my own class": novice language teacher identity construction in Turkey. *Asia Pacific Journal of Education*, 1–16.

Gün, B. (2011). Quality self-reflection through reflection training. *ELT Journal, 65*(2), 126–135.

Higgins, C., & Ponte, E. (2017). Legitimating multilingual teacher identities in the mainstream classroom. *The Modern Language Journal, 101*(S1), 15–28.

Hong, J., Greene, B., & Lowery, J. (2017). Multiple dimensions of teacher identity development from pre-service to early years of teaching: A longitudinal study. *Journal of Education for Teaching, 43*(1), 84–98.

Huang, X., Lee, J. C. K., & Yang, X. (2019). What really counts? Investigating the effects of creative role identity and self-efficacy on teachers' attitudes towards the implementation of teaching for creativity. *Teaching and Teacher Education, 84*, 57–65.

Jiang, L., Yuan, K., & Yu, S. (2020). Transitioning from Pre-service to Novice: A Study on Macau EFL Teachers' Identity Change. *The Asia-Pacific Education Researcher*, 1–11.

Kayi-Aydar, H. (2019). Language teacher identity. *Language Teaching*, 52(3), 281–295.

Lave, J., & Wenger, E. (1991). *Situated learning: Legitimate peripheral participation*. Cambridge, UK: Cambridge University Press.

Li, W. (2020). Unpacking the complexities of teacher identity: Narratives of two Chinese teachers of English in China. *Language Teaching Research*, 0(00), 1–19.

Ruohotie-Lyhty, M., & Pitkänen-Huhta, A. (2020). Status versus nature of work: pre-service language teachers envisioning their future profession. *European Journal of Teacher Education*, 1–20.

Sang, Y. (2020). Research of Language Teacher Identity: Status Quo and Future Directions. *RELC Journal*, 0(00), 1–8.

Vygotsky, L. S. (1978). Socio-cultural theory. *Mind in society, 6*, 52–58.

Yuan, R. (2019). A critical review on nonnative English teacher identity research: From 2008 to 2017. *Journal of multilingual and multicultural development, 40*(6), 518–537.

Yuan, R., Liu, W., & Lee, I. (2019). Confrontation, negotiation, and agency: exploring the inner dynamics of student-teacher identity transformation during teaching prácticum. *Teachers and Teaching, 25*(8), 972–993.

Post Pandemic L2 Pedagogy – Adi Putra & Arifah Drajati (Eds)
© 2021 Taylor & Francis Group, London, ISBN 978-1-032-05807-8

Developing an electronic pocket dictionary based on the ADDIE model for Bahasa Indonesia basic learners

M. Fernanda Adi Pradana & Putri Kumala Dewi
Universitas Brawijaya, Indonesia

ABSTRACT: This research focuses on the development of Indonesian language learning tools for Bahasa Indonesia basic learners. Therefore, it aims to develop easy access tools that basic Indonesian learners can use to hear in Bahasa Indonesian. This research used the ADDIE method, which stands for Analysis, Design, Development, Implementation and Evaluation, to create the targeted product named Electronic Pocket Dictionary for Bahasa Indonesia basic learners. The development of an Electronic Pocket Dictionary based is focused on the feasibility of content, language and systematic presentation. The researchers interviewed linguists, media experts, practitioners and learners from one of the universities in Indonesia to get their desired results. The results show that the Electronic Pocket Dictionary meets a good category according to validations from material experts, linguists, media experts, practitioners and learners. In addition, the researchers found that it is not an easy thing to create a learning medium, but necessary so that the field of education does not use obsolete technologies.

Keywords: ADDIE Model, Bahasa Indonesia Basic Learners, electronic dictionary

1 INTRODUCTION

Bahasa Indonesia has an important position as a foreign language globally as the Indonesian government wishes that it can be a language that people in the world use as a second foreign language. Therefore, BIPA is a program of the Indonesian government to globalize Bahasa Indonesian by teaching Bahasa Indonesian to foreigners. Quoted from the page www.bipa.kemendikbud.id, the BIPA program is spread across various regions globally, including the United States, Australia, Europe, Africa and the Asian Pacific. Bahasa Indonesian's learning effort for foreign speakers requires the Indonesian government to facilitate learning modules. The government has created the book entitled *Sahabatku Indonesia* as a learning resource for Indonesian basic learner students published in 2016 by the Ministry of Education and Culture of the Republic of Indonesia. However, *Sahabatku Indonesia,* which has been used as a teaching module for Indonesian basic learners, still has weaknesses such as a lack of Indonesian vocabulary and sentence structures that still confuse students, writing errors and a vocabulary that is complicated for Basic Learners to learn Bahasa Indonesian. Yet, the learning media still has some issues that can be adjusted, such as,the reading skills comprehension section is not based on daily issues; the text is still too long and several structures confuse Indonesian basic learners. Second, the listening skills learning process uses many incomprehensible words that are used in the whole text, and some words have no reference for daily use. Third, in writing skills, there are still errors in writing the letters l and r in the words for give and buy (membeli dan memberi), -ng in the word count (menghitung), and differentiating the use of affixes. Fourth, speaking skills have not been accommodated by learners to speak skillfully and apply them every day. Hence, it is important to evaluate *Sahabatku Indonesia* and improve it to become a more suitable learning tool for them.

The researchers use some previous studies as references. First, it is a study by Durak and Ataizi (2016), which analyzed the development of programming language learning media for

DOI 10.1201/9781003199267-13

undergraduate students utilizing ADDIE. This previous study aims to develop media for designing an online course for programming language. Second, it is research done by Ghani and Daud (2018). The researchers aim to analyze website learning media for Arabic for Tourism Purpose lessons using the ADDIE model and develop it into a better media for learning Arabic. Moreover, the previous researchers also intends to give a new experience, improve learning materials and make the learning process more effective. It is in line with the final research by Saman et al. (2019) as the third reference. It analyzed the game's development as a learning media for people with a hearing problem using the ADDIE model. This previous study results in a product of a game for learning sign language, named I-Sign. In all of the previous studies, it can be said that learning media should be improved and developed so that the learners do not get jaded by the materials and the digital generations have the interest to learn the materials.

The previous studies mentioned above using the ADDIE model to develop learning media. Yet, none of them apply the model to develop a dictionary for foreign learners. Moreover, Durak and Ataizi (2016) and Ghani and Daud (2018) use the model to develop learning tools that have already existed before, meanwhile, Saman et al. (2019) make their own product as the result of analyzing people with disability's needs. Even though the Instructional Design Model used is the same, the objectives are not similar in finding out the learning media users' needs. The three previous studies use different methods, one of which uses questionnaires while the two others use informal interviews with both teachers and students. After reviewing the previous studies, the current researchers decided to develop a new application for Basic Learners in Learning Bahasa Indonesia who need more updated learning media and interview them to know what their specific needs are, so the learners can use them in their daily lives. The present research aims to develop the electronic pocket dictionary for the basic learner's student of Bahasa Indonesia. The research limits on the nine Bahasa Indonesia Basic students from Japan, Cambodia, South Korea, Egypt, Thailand, Tunisia and Vietnam. The students learn basic Bahasa Indonesian in Universitas Brawijaya. The developing media of electronic pocket dictionaries hopefully can be useful for the learners to learn Bahasa Indonesia as a second language. This research is expected to make it easy for foreign students to learn Bahasa Indonesian with effective devices.

2 LITERATURE REVIEW

Muhdi et al. (2019), learning media is a physical instrument, both hardware and software, so that it can stimulate thoughts, feelings, attention, and student's interests and concern for the learning process to occur. Aqib (2014) adds that the meaning of media learning is delivering information, stimulating the mind, and encouraging an interesting learning process and learning objectives are achieved easily. There are four stages in this experiential strategy, including real experiences, individual observation and reflection, conceptualization, and implementation (Chiu 2019). This communicative approach is an approach that is based on Indonesian Basic Learners to master the Indonesian language and can be used in daily and informal context communication activities (Issakova 2018).

Ellsa and Rahmawati (2020) researched the Development of Indonesian Language Teaching Materials for Bahasa Indonesia Basic Learners. This research's output is a learning media in Word Cards, which is based on the Graduate Competency Standard in Indonesia. It is the same as the current study. This research found that traditional learning material gives basic learners difficulties understanding the material since they become passive learners. The second study regarding the development of learning media for Indonesia's basic learners is Lestari et al. (2019). This research produces flashcards as the learning media. Unlike previous studies, the current study develops learning media through an application using Artificial Intelligence (AI).

From the previous studies used as a reference, the utilization of ADDIE to develop learning and information tools for various people is the most prominent similarity. Meanwhile, the differences are certainly in the language subject taught through the products produced. They are namely literacy courses for children, communicative English for employees, and Turkish for foreign learners (Hess & Gree 2016; Karademir et al. 2019; Rafiq et al. 2019). Nevertheless, the object of this study is the Indonesian language learned by Basic Learners in Learning Indonesian. The previous study shows that computerized book media improved students' citizenship instruction understanding capacities more successfully than would happen within customary learning.

3 RESEARCH METHOD

This study's participants were nine Bahasa Indonesian Basic students (from Japan, Cambodia, South Korea, Egypt, Thailand, Tunisia and Vietnam). The primary data is the result of interviews with Bahasa Indonesian basic learners students in Universitas Brawijaya on September 24, 2019. This interview aims to find and determine learners' needs in learning Bahasa Indonesian. After that, various experts (material, language, and media) and Bahasa Indonesia Learners practitioners on electronic pocket dictionaries validate the data which comes from the development of each product until the final one is categorized as good or well. The second, the trial results of Bahasa Indonesia Basic learners in Malang on the electronic pocket dictionary as this application is made for them.

The type of research used is a design-based research model. Design-based research is claimed to have the potential to bridge the gap between educational practice and theory because it aims both at developing theories about domain-specific learning and the means that are designed to support that learning process (Bakker & Van Eerde in press). This research develops media innovation in electronic pocket dictionaries based on android, which is used as a daily supplement for Bahasa Indonesian basic learners in Malang. The ADDIE model is the research model used in this study. It contains five stages of development: analysis, design, development, implementation, and evaluation (Branch in Hess & Greer, 2016). The ADDIE model provides a structured and interrelated process so that its implementation must be structured and cannot be arranged randomly.

3.1 *About electronic pocket dictionary*

Electronic Pocket Dictionary named E-KAMUSKU stands for '*Kamus Saku Elektronik.*' an electronic pocket dictionary is an Android-based multimedia type and is operated via smartphone. This application was developed using the Android Studio application. It carries a simple theme and language so that Basic Learners in Learning Indonesian can quickly understand this application. The color used in the electronic pocket dictionary application is a combination of red and white, which is the Indonesian flag's colors. Another advantage is that this media contains much Indonesian vocabulary, which is differentiated by various themes based on the objectives contained in ACTFL and includes four language skills: reading, listening, writing and speaking, which can make it easier to sharpen the language proficiency.

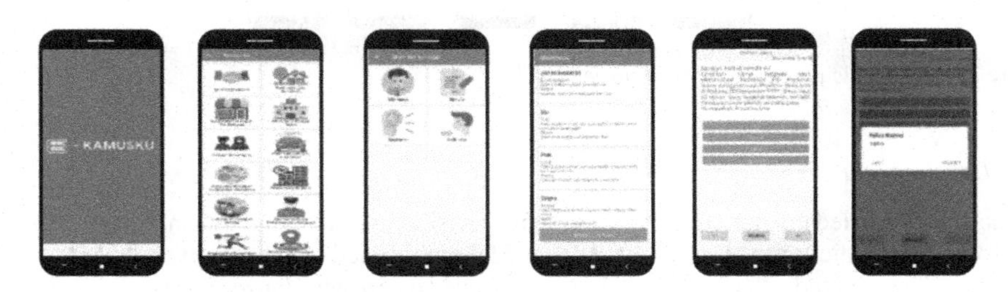

Figure 1. System design of electronic pocket dictionary.

4 FINDINGS AND DISCUSSION

The results of this development will be discussed regarding the description of research procedures with the ADDIE model, including analysis, design, development, implementation, evaluation.

4.1 *Analysis*

The results obtained are about three things, the results of the researchers' observations on the Sahabatku Indonesia book, which found problems regarding the content related to language skills, i.e., reading, listening, writing and speaking. Potential analysis related to Bahasa Indonesian basic students from various countries, including Japan, Cambodia, South Korea, Egypt, Thailand, Tunisia and Vietnam, has different characteristics. Finally, a needs analysis related to interviews with nine students showed that Bahasa Indonesian basic learners need to find supplements that are used to develop communication in reality that are fast, portable and easily accessible wherever they are. Language learners have to learn frequently to acquire skills faster and better (Nation & Yamamoto 2012). Besides, they also need learning media that teaches all aspects of language. This statement is in line with Nation's (2007) statement which states, "a well-balanced language course should consist of four equal strands – meaning-focused input, meaning-focused output, language-focused learning, and fluency development. Each strand should receive a roughly equal amount of time in a course." So, a media that includes every aspect or way of learning each language skill is needed by Basic Learners in Learning Indonesian.

4.2 *Design*

The results obtained are about three things, namely electronic pocket dictionary navigation, flowcharts and storyboards. The navigation design is related to creating an overview of the menus in the applications and direction patterns. Furthermore, its flowchart design is related to the flow description included in this media. The flowchart flow in an electronic pocket dictionary is designed as follows. Additionally, the flowchart relates to the electronic pocket dictionary storyboard design related to the sketch or display picture and scenario or what things are done on each electronic pocket dictionary page. The storyboard description in its media is presented below.

Figure 2. Electronic pocket dictionary storyboard.

4.3 *Development*

The activities carried out are creating, validating, revising product I and assembling product II. An electronic pocket dictionary is developed using 12 themes based on the real experiences of learners in Malang. They include, "*Let's Get Acquainted, Come Over to My House, Shop at Traditional Markets, Go to Hospital, Study on Campus, Ride Indonesian Transportation, Try Indonesian Traditional*

Cuisine, Save Money at the Bank, Visit Tourist Attractions, Various Jobs in Indonesia, My Daily Activities, and *Understanding Directions.*" Then there are 315 vocabularies and 108 questions. An electronic pocket dictionary also includes four English skills, which are reading, listening, writing, and speaking.

The product development is inseparable from suggestions and comments regarding the appropriateness of content, language, media and trials to Bahasa Indonesian learners. The results of the validations, comments, and suggestions for the product I are described as follows. From the results of the first validation, the eligibility of the content got a value of 71% in the Good category, language eligibility got a score of 53% in the Bad category, the media feasibility got a score of 89% in the Excellent category, and Bahasa Indonesian learners got a score of 85% in the Excellent category. These first validation results produce the following suggestions and comments in the suitability between contents and learning objectives, material accuracy, diction, communicative, and serving support. Finally, from presenting the suggestions and comments from Bahasa Indonesian learning experts and students, the product was revised. Product I, which has been revised, is called product II. In this phase, practitioners' reviews are crucial since the researchers need feedback before the product is given to the users. Furthermore, the tools that are used in the design phase are still utilized in this phase, as it is similar to Durak and Ataizi(2016), Ghani and Daud (2018), and Saman et al. (2019).

4.4 *Implementation*

At the implementation stage, the activities carried out include product validation II, small-scale trials, revision II, product III and wide-scale trials. The validation results, comments, and suggestions for product II are described as follows:

From the results of this second validation, it can be seen that the eligibility of the content got a score of 80% in the Good category, the feasibility of the language got a score of 73% in the Good category, the eligibility of the media got a score of 100% in the Excellent category. BIPA practitioners got a score of 91% in the Excellent category. The second validation summary results in the following suggestions and comments on the suitability between content and learning objectives.

Next, a limited trial was carried out for four Bahasa Indonesia Basic Learners from one of the universities in Indonesia. The students come from Japan, South Korea, Thailand and Tunisia. Learners are selected randomly or usually known as random sampling. The results obtained are as follows. This limited trial's implementation concludes that it scores an average of 91% in the Excellent category, and no suggestions or comments were found on product II. Therefore, the next step is to revise product II from the suggestions and comments on the content's feasibility to develop product III. Product III is ready to be implemented to nine Bahasa Indonesia basic learners. The results obtained are as follows.

The implementation of this extensive trial resulted in a scoring average of 91% in the Excellent category, and no suggestions and comments were found on product III. Therefore, product III is ready to be packaged to produce the final product. Since this phase allows all materials to be tested to identify if they are well-functioning and appropriate for the intended audience (Ghani & Daud, 2018), the researchers then put it online and shared it with the users,

4.5 *Evaluation*

There are no revisions to product III at this stage, so the packaging process of the electronic pocket dictionary media is carried out. The packaging is done by distributing it through Google Drive. The Google drive address of the application is at bit.ly/E-KAMUSKU.

4.5.1 *Media studies of electronic pocket dictionary*

The findings are related to the use of experiential strategies and communicative approaches associated with the Graduate Learning Outcomes (CPL) of Bahasa Indonesia Basic Learners. An electronic pocket dictionary can fulfill the CPL Bahasa Indonesia Basic Learners in reading and listening skills in content feasibility. The reading skills are proven by selecting words and texts in vocabulary and evaluation using daily use themes. The selection of these themes is closely related to the early stages of experiential strategies, namely, real experiences. The listening skills are then proven by choosing words, phrase and simple sentences with references to reality since it is evidenced by experience, which focuses on the contextual arrangement of vocabulary.

An electronic pocket dictionary can meet CPL Bahasa Indonesian basic learners' standards in reading and writing skills in the aspect of language eligibility. Reading skills are proven by selecting text in the evaluation using 4–5 sentences. The scoring is related to Bahasa Indonesia's basic learners' readability level, which is only introducing a limited number of letters, symbols and characters. Furthermore, the use of single sentence structures in sentences and evaluation so that learners do not experience errors in understanding the reading. Then, in writing skills, it is proven by the use of speaking and example sentences in the vocabulary so that learners do not experience errors in writing the letters l and r in words *memberi* and *membeli* (give and buy), writing the letter -ng in the word *menghitung* (counting) and also the use of affixes.

In the aspect of media feasibility, the electronic pocket dictionaries can fulfill the CPL Bahasa Indonesia Basic Learners in speaking skills. The scoring is proven by the facilities in electronic pocket dictionaries that can make learners practice speaking. The content is provided to know the correct words, phrases and sentences pronounced, or there are still pronunciation errors. This facility is related to the communicative approach, which is the ultimate goal of using an electronic pocket dictionary; learners can master the Indonesian language and communicate daily.

Therefore, from the results of evaluations from the experts and users, it can be said that the development of this language learning tool has succeeded. This statement is also supported by the fact that the whole process in developing an electronic pocket dictionary fulfills the user's need to learn Bahasa Indonesia, in this case, basic learners in learning Indonesian. Moreover, each step of the ADDIE Model is also done based on the initial analysis at the beginning of the research. In this research, the researchers have analyzed the user's needs based on the users' skills, the model of the tools, the objectives, and the learning tool's goals (Aldoobie 2015). After that, the researchers designed the product, which took some time and a couple of revisions to meet the objectives and the users' expectations, which went right through to the development phase. So, in those phases, the users' feedback is essential. This statement is in line with Aldoobie's (2015) statement, which says that users are expected to be involved in the use of strategies during the design and development phases and give feedback regarding the products. Finally, the final product is given to some experts and users so they can use and evaluate the product to determine whether the product is qualified or not.

5 CONCLUSION

Based on the results of development and small-scale and large-scale trials related to electronic pocket dictionary media as supplementary materials for Bahasa Indonesian basic learners, several things can be concluded from these media. There are elements of communication and vocabulary that are easy to access wherever you are. There are references (how to speak and sample sentences) of the vocabulary provided while accommodating the actual context theme. Moreover, the results of this study are supported by the previous studies relating to the learning tools ADDIE Model development, which show that by learning using the developed learning tools, learners will grasp the materials easier (Durak & Ataizi 2016; Ghani et al. 2018; Wibawa et al. 2017). Unfortunately, this study has not covered all the basic learners in learning Indonesian since it only focuses on basic learners in learning Indonesian. Therefore, the researcher expects future researchers to develop this program to cover every Bahasa Indonesia Basic Learner's needs from all learner levels. The researchers also hope for future researchers to improve this pocket dictionary application. They are applications only compatible with Android-based phones; pronunciation does not voice out loud, the number of examinations is limited, and the application is not available on the PlayStore. During the research and through the findings, the researchers found that developing a learning media is not as easy as Mohan states (2001) and that learning media should give impacts on 1) the effectiveness of time, 2) student's interests, 3) student's attention, 4) the clarity of idea, 5) the strength of the concept taught, 6) the teaching tone, 7) the point made through the media, and 8) student's memory.

REFERENCES

Aqib, Z. (2014). *Model-model, media, strategi pembelajaran kontekstual (inovatif)*. Bandung: Yrama Widya.

Bakker, A., & Van Eerde, H. A. A. (in press). An introduction to design-based research with an example from statistics education. In A. Bikner-Ahsbahs, C. Knipping, & N. Presmeg (Eds.), Doing qualitative research: methodology and methods in mathematics education. New York: *Springer*.

Chiu, S. K. & Lee, J. (Reviewing editor). 2019. Innovative experiential learning experience: Pedagogical Adopting Kolb's Learning Cycle at Higher Education in Hong Kong. *Cogent Education* 6(1).

Durak, G., & Ataizi, M. (2016). The ABC's of online course design according to addie model. *Universal Journal of Educational Research*, 4(9), 2084–2091.

Ellsa, S & Rahmawati, L. (2020). Pengembangan media kartu kata dalam pembelajaran bahasa Indonesia bagi penutur asing. *SAP (Susunan Artikel Pendidikan)*.

Ghani, A. Taufiq, M. & Daud, W. (2018). Adaptation of addie instructional model in developing an educational website for language learning. *Global Journal Al-Thaqafah*. 8, 7–16.

Hess, K. A. & Greer, K. Designing for engagement: Using the ADDIE model to integrate high-impact practices into an online information literacy course. *Communications in Information Literacy* 10(2) 264–282.

Issakova, A. (2018). Communicative. approach to interactive foreign language lesson at University. *Cildiah,* 50. 1–5.

Karademir, T. ğ., Alper, A., Sğuksu, A. F., & Karababa, Z. C. (2019). The development and evaluation of self-directed digital learning material development platforms for foreign language education. Interactive Learning Environments, 1–18.

Lestari, N. M. C. P., Sutama, I. M., & Utama, I. D. G. B. (2019). Pengembangan media pembelajaran visual bagi pembelajar BIPA pemula di UNDIKSHA. *Jurnal Pendidikan Bahasa Dan Sastra Indonesia Undiksha, 8*(1), 86–95.

Mohan, T. (2001). *Communicating Theory & Practice*. Thomson Reuters.

Muhdi, et al. (2019). Design of digital book media to teach citizenship education with a contextual approach. *Journal of International Journal of Innovation, Creativity, and Change 9(10).*

Muslimin, Siri Mohammad, et al. (2017). The Design and Development of Mobieko: a Mobile Educational App for Microeconomics Module. *Malaysian Journal of Learning and Instruction* 221–255.

Nation, P. & Yamamoto, A. (2012). Applying the four strands to language learning. *International Journal of Innovation in English Language Teaching,* 1(2).

Rafiq, M., Hashim, H., Yunus, M., & Pazilah, F. (2019). Developing a MOOC for communicative English: A Battle of instructional designs. *International Journal of Innovation, Creativity, and Change, 7(7), 29–39.*

Richey, R. C. & Klein, J. D. (2005). Developmental research methods: Creating knowledge from instructional design and development practice. *Journal of Computing in Higher Education* 16(2). 23–38.

Saman, F., Shariff, N. & Nasaruddin, N. I. S. (2019). i-Sign: Sign language learning application via gamification. *Asian Journal of University Education*. 15.

Solikhah, I. and Budiharso, T. (2020). Standardizing BIPA as an international program of a language policy. *The Asian ESP Journal* 16(5.2) 166–190.

Toro, V. & Camacho-Minuche, G. & Pinza, E. & Paredes, F. (2018). The use of the communicative language teaching approach to improve students' oral skills. *English Language Teaching* 12. 110–118.

Post Pandemic L2 Pedagogy – Adi Putra & Arifah Drajati (Eds)
 ISBN 978-1-032-05807-8

Readiness for technology-based teaching among prospective English teachers in Indonesian border universities

Lita Liviani Taopan, Christi A. Malaikosa, Imanuel Y.H. Manapa, Santhy Givend Pandie & Agustina Aloojaha
Universitas Tribuana Kalabahi, East Nusa Tenggara, Indonesia

ABSTRACT: Technology-based teaching is announcing novel pedagogical methods or emerging ones using Information and Communication Technology (ICT) to support modern electronic media to fulfill educational constraints. This study intended to reveal the prospective teacher's readiness for English education in a border university regarding the technology integration in teaching and to reveal the barriers. In this study, prospective teachers' readiness was assessed under technological competency, pedagogical competence, and technological pedagogical competency developed based on the TPACK framework developed by Mishra & Koehler (2006). A descriptive quantitative was done at the English Department of universities located in the border area of Indonesia. Forty-eight prospective English teachers who participated in this study reveal that they are not fully ready for technology integration in the future, yet they have a strong desire to learn about it. The students' main obstacles to learning with technology were the facilities and the teacher's capability from the previous academic level. As the preliminary study, this study's result would be beneficial for future research.

Keywords: Border area, prospective English teacher, readiness, TPACK.

1 INTRODUCTION

The integration of Information and Communication Technology (ICT) has been an essential agenda of all countries globally due to the massive and rapid changes occurring now and then. It connects every part of the world. It has become a critical tool to keep up with the rapid advancement in the recent era known as the 21st century. Thus, such an extent should be seriously taken into account by the countries (including Indonesia) not to be left out. With no exception, all citizens should be aware of ICT skills and knowledge to get through the ever-changing world.

Additionally, Information and Communication Technology (ICT) has been increasingly prominent in the educational world. With the help of technology, the world can be presented in a classroom. The pivotal role of ICT is nuanced in 21st-century skills that both contemporary and future students expect to get from schools (Valtonen et al. 2018). All of the skills, readiness for collaboration, problem-solving, and creative & critical thinking, are the most fundamental, and ICT is seen as a means of teaching the skills mentioned above (Voogt & Roblin 2012). Furthermore, ICT helps students to interact with the rest of the world easily. This situation allows students to access complete knowledge and information. Students can interact and engage in collaborative learning with peers from around the world regardless of time and space.

To cope with the situation, teaching approaches and teachers' roles should be readjusted to cater to such a new way of teaching and learning process. Teachers should be able to adapt to changes, and most importantly, they should be aware of this exigency during the whole of their pre-service teacher training (Strakova 2015). In other words, all responsible stakeholders need to take advantage of this situation to create a technology-enhanced learning environment. Despite teachers' willingness and interest to integrate ICT into the educational environments in their various

 DOI 10.1201/9781003199267-14

institutions, many are constrained by a lack of confidence, training, skills, technical support, resources as well as technological infrastructure to carry it out effectively (Krause et al. 2017). Accordingly, teacher education is responsible for providing teachers with readiness to integrate ICT for various learning purposes. The expectations for preparing prospective teachers to use ICT are considered as twofold. Firstly, pre-service teachers are a part of the digital natives who actively utilize ICT in their everyday lives for educational purposes (Tapscott 2008). Secondly, today's prospective teachers are the generation who extensively use numerous applications but still lack integrating the ICT for educational purposes (Lei 2009).

Although the readiness for technology-based teaching among prospective teachers is undoubtedly substantial, assessing the readiness for technology-based teaching of these future educators seems overlooked. This study intended to reveal the readiness of the prospective teachers of English education in the border universities regarding the technology integration in teaching and to reveal the barriers. In this study, the readiness of pre-service teachers was assessed under the TPACK (technological competency, pedagogical competence, and technological pedagogical competence) framework by Mishra and Koehler (2006).

2 LITERATURE REVIEW

There have been various techniques utilized to assess the readiness for integrating technology for educational purposes based on their suitability toward the assessed domains. For this study, the TPACK framework was used to assess the readiness of the prospective teachers.

Technological, Pedagogical, Content Knowledge (TPCK) was first announced by Mishra and Koehler (2006) as a comprehensive framework of teachers' knowledge for integrating technology in education. Then, the term was renamed Technological, Pedagogical, and Content Knowledge (TPACK). Koehler and Mishra (2009) defined the required bodies of knowledge constituting the basis of the TPACK framework for an effective ICT integration into the teaching and learning process. The bodies encompass technological knowledge, content knowledge and pedagogical knowledge. Further, the interaction between the three core bodies is the essence of TPACK. This framework postulates that to integrate ICT into education effectively, it is irrevocable for teachers to understand the interaction and the interrelation between technology, pedagogy, and content to construct a form of knowledge beyond these three separate knowledge bases (Harris et al. 2009).

Precisely, the integration of technology into the three core types of knowledge culminated in three additional types of knowledge; Pedagogical content knowledge (PCK), Technological content knowledge (TCK) and Technological pedagogical knowledge (TPK). Then, these three bodies of knowledge's interaction result in TPACK (Graham 2011; Hechter et al. 2012). As teachers are required to possess the knowledge of pedagogy, content and technology and how these three foundations can be engaged interdependently in teaching for educational goals, the TPACK framework, therefore, is considered suitable for assessing teachers' readiness to equip ICT for teaching and learning effectively.

A massive number of studies present the use of the TPACK framework to assess teachers' readiness or ability regarding ICT integration into teaching and learning. Firstly, Valtonen et al. (2018) conducted a study regarding the distinctions in teachers' knowledge and readiness to integrate ICT into education. The study revealed some differences based on technological, pedagogical, and content knowledge (TPACK) and the area of theory of planned behavior (TPB). The study also suggested carefully understanding teachers' different capabilities to provide them with the support they need.

Secondly, Gyaase et al. (2019) conducted a study to assess the extent of prospective teachers' e-readiness to utilize ICT in teaching various subjects. By using the TPACK framework, their study revealed that the participants' ability to engage ICT to design and deliver the subject contents as well as to improve the learning atmosphere was still low, even though they possess high ICT literacy. Further, the study suggested a more comprehensive preparation of teachers with capabilities to use ICT in their teaching. This study's objective was to discover the level of readiness for technology-based teaching among prospective English teachers in Indonesian border universities.

3 METHOD

This study used a descriptive research method with a quantitative approach. The data was collected through questionnaires and online interviews. Descriptive research usually uses only one variable, so it tends not to be intended to reveal relationships between variables or test hypotheses. In this study, 48 prospective teacher students were the study population. A sampling method was not used. The whole population was considered as the sample since the whole population was accessible. The quantitative data was collected through a self-administered questionnaire, and the data were analyzed descriptively using SPSS: version 22. The questionnaire was a modified questionnaire based on the instrument developed by Scmith (2009), Consisting of seven domains of the TPACK framework and used a seven Likert scale since it offers seven different answer options related to an agreement that would be distinct enough for the respondents, without throwing them into confusion.

4 RESULT AND DISCUSSION

The statistical description of the data in this study is intended to reveal the readiness of technological pedagogical content knowledge (TPACK) of prospective English teachers at the universities in the Indonesian Border Area. The result was derived from the frequency analysis using SPSS 22. In detail, the results of the data analysis for each domain of TPACK are presented in the following description:

Firstly, Technological Knowledge (TK). This is knowledge related to technology and its application (Jordan 2011). The results of the descriptive analysis show that the level of readiness for the Technological Knowledge (TK) of the prospective English teacher are: the mean is 5.25, the middle value (median) 5,78, the value that often appears or the mode 5.85, and the standard deviation (SD) 1.321. Meanwhile, the highest percentage of responses related to the TK readiness is on the term "agree," (37.5%). It can be concluded that almost half of the English study program students agree that they have already mastered the fundamental knowledge of technology. Of the 48 respondents, only 4.2% of students stated that they did not have technology-related knowledge integrated into teaching English.

Secondly, Content Knowledge (CK). It deals with knowledge about concepts, theories, ideas, frameworks, knowledge about proof, and practices and approaches to developing this knowledge (Shulman 1986). For the readiness of the content knowledge: the mean 6,03, the median 6, mode 6, and the standard deviation (SD) 0.94. Meanwhile, the analysis of the readiness for content Knowledge shows the highest percentage on the term "agree" (55.5%). It means that more than half of the respondents agree that they have sufficient content knowledge, which is, in this case, English language knowledge.

Thirdly, Pedagogical Knowledge (PK) is the teacher's knowledge of various implementations, strategies, and methods to support student learning (Koehler et al. 2013). For the readiness of the pedagogical knowledge: the mean 5,11, the middle value (median) 5,35, the mode 5,57, and the standard deviation (SD) 1,56. For the analysis results, the Content Knowledge item got the highest percentage on the term "agree" (29,7%). However, the percentage is not to reach half of the respondents. While the percentage of "neutral" is 20%. It indicates that not all prospective teachers in the border area agree that they have to master sufficient knowledge about various teaching methods.

Fourthly, Pedagogical Content Knowledge (PCK) is pedagogical knowledge applicable to teaching specific content. This knowledge includes knowing what teaching approaches are appropriate to the content and how content elements can be arranged for better teaching (Mishra &Koehler 2006).

Based on descriptive analysis for the pedagogical content knowledge's readiness: the mean 5,25and the standard deviation (SD) 1,42. The analysis results for the readiness of content Knowledge show the highest percentage on the term "agree" (33,3%). It indicates that most prospective English teachers do not have sufficient knowledge about various teaching methods to teach English.

The next, Technological Content Knowledge (TCK) is knowledge about the reciprocity between technology and content (Koehler 2014). The descriptive analysis for the readiness of the technological content knowledge shows a mean of 5.5. the middle value (median) 6, the value that often appears or the mode 7, and the standard deviation (SD) 1.42. Moreover, the analysis of the readiness of technological content knowledge shows the highest percentage on the term "strongly agree" (31,3%). It indicates that most prospective English teachers in Indonesia's border area universities are ready to implement the technology for teaching the specific content.

The sixth, Technological Pedagogical Knowledge (TPK) is about how various technologies can be used in teaching (Koehler 2014). For the content knowledge's readiness, the mean 5,49, the middle value (median) 6, the value that often appears or mode 6, and the standard deviation (SD) 1.48. The analysis results for the readiness of technological content Knowledge show the highest percentage on the term "agree" (41,26%). It indicates that most prospective English teachers are ready to integrate technology with various teaching methods.

Finally, Technological Pedagogical and Content Knowledge (TPACK) is the knowledge needed by teachers to integrate technology into teaching. However, teachers need to understand the complex interactions between the three essential components of knowledge, namely PK, CK, and TK, by teaching certain materials using appropriate pedagogical and technological methods (Schmidt et al. 2009). The descriptive statistic analysis for the readiness of the Technology, Pedagogical and Content Knowledge, the minimum score was 2, the maximum score was 7, the mean was 5,27, the middle value (median) was 6, the value that often appears or the is mode 6.2, and the standard deviation (SD) was 1.53. Meanwhile, the analysis of the readiness of content Knowledge shows the highest percentage on the term "agree" (35%).

Apart from the seven domains of the TPACK, we also add questions related to the teachers' ability and support facilities. The result shows that 80% of respondents strongly agree that the universities have lecturers that teach the technology subject. Moreover, 75% of respondents agree that teachers are capable of integrating technology in teaching. They also agree that the teachers are good role models in integrating technology in teaching. Furthermore, related to the facilities, we cross to a different reality where only 20% agree that the university facilities are sufficient to support the use of technology. Such as the internet connection and the facilities such as a computer or other devices. It is a general problem faced by the universities in the rural area, including Indonesia's border area.

5 CONCLUSION AND RECOMMENDATION

Based on the description above, several conclusions can be drawn. First, students in Indonesia's border areas have great potential and a strong desire to integrate technology in teaching. Most respondents have declared their readiness for technology integration. It could be seen in the frequency of responses, in which the average of the highest presentations was on the item "agree." It means that the prospective English teachers in the border area of Indonesia realize that they cannot avoid technology integration in their future careers as English teachers. Therefore they have to prepare themselves to be ready for future changes. Second, the ability of the lecturers of the universities in the border area of Indonesia for integrating technology in teaching is good. Most of the lecturers are capable of using the latest technology in teaching. The last, the only problem that still becomes the main concern is the availability of facilities to support technology integration.

Furthermore, the findings of this study could be the recommendation for the stakeholders in the border area of Indonesia to have more concern about the supporting facilities in the universities. The leader of universities should be aware of this problem and take action to support education in the universities in the border area of Indonesia.

REFERENCES

Graham, C. R. (2011). Theoretical Considerations for Understanding Technological Pedagogical Content Knowledge (TPACK). *Computers & Education, 57*(3), 1953–1969.

Gyaase, et al. (2019). Gauging the E-readiness for the Integration of Information and Communication Technology Into Pre-Tertiary Education in Ghana: An Assessment of Teachers' Technological, Pedagogical, and Content Knowledge (TPACK). *International Journal of Information and Communication Technology Education. 15*(2).

Harris, J., Mishra, P., & Koehler, M. (2009). Teachers' Technological Pedagogical Content Knowledge and Learning Activity Types: Curriculum-based Technology Integration Reframed. *Journal of Research on Technology in Education, 41*(4), 393–416.

Hechter, R. P., Phyfe, L. D., & Vermette, L. A. (2012). Integrating technology in education: Moving the TPCK framework towards practical applications. *Education Research and Perspectives, 39*(1), 136–152.

Jordan, K. 2011. *Beginning Teacher Knowledge: Result from Self-Assessed TPACK Survey*. Australian Educational Computing. 26 (1)

Koehler, M. J., & Mishra, P. (2009). What is technological pedagogical content knowledge? *Contemporary Issues in Technology & Teacher Education, 9*(1), 60–70.

Koehler, M.J., Mishra, P., Akcaoglu, M., & Rosenberg, J.M. 2013. *The Technological Pedagogical Content Knowledge Framework for Teachers and Teacher Educators. ICT Integrated Teacher Education.*

Krause, M., Pietzner, V., Dori, Y. J., & Eilks, I. (2017). Differences and Developments in Attitudes and Self-Efficacy of Prospective Chemistry Teachers Concerning the Use of ICT in Education. *Eurasia Journal of Mathematics. Science and Technology Education, 13*(8), 4405–4417.

Lei, J. (2009). Digital natives as preservice teachers: What technology preparation is needed? *Journal of Computing in Teacher Education, 25*(3), 87–97.

Mishra, P., & Koehler, M. J. (2006). Technological pedagogical content knowledge: A framework for teacher knowledge. *Teachers College Record, 108*(6), 10171–10154.

Tapscott, D. (2008). *Grown-up digital: How the net generation is changing your world HC*. New York, NY: McGraw-Hill.

Valtonen, et al. (2018). Differences in pre-service teachers' knowledge and readiness to use ICT in education. *Journal of Computer Assisted Learning*. 34:174–182.

Voogt, J., & Roblin, N. P. (2012). A comparative analysis of international frameworks for 21st-century competencies: Implications for national curriculum policies. *Journal of Curriculum Studies, 44*, 299–321.

Post Pandemic L2 Pedagogy – Adi Putra & Arifah Drajati (Eds)
 ISBN 978-1-032-05807-8

Correlation between students' perception of blended learning and their learning outcomes

Juliaans E.R. Marantika & Jolanda Tomasouw
German Language Education Department, Universitas Pattimura, Indonesia

ABSTRACT: This study was conducted to determine students' perceptions of the blended learning model. It used a quantitative research method to analyze the correlation between students' perception and learning outcome of the Aufbaustufe course and conducted for three months. The participants of this research consist of 32 students of the German Language Education Department from the third semester of the 2019/2020 academic year. Data were collected by questionnaires to assess students' perceptions of blended learning and tests to measure Aufbaustufe's learning outcomes. Data were analyzed using the Pearson correlation technique. The result implied that the better the student's perception of the blended learning model, the better the student learning outcomes in the Aufbaustufe course. Therefore, modeling a good learning strategy can change the perceptions of students, which improves learning achievement.

Keywords: Blended Learning, German Language Teaching

1 INTRODUCTION

Trilling and Fadel (2012) identified three qualifications that are expected from students in 21st-century learning, namely: (1) Learning and innovation skills; (2) Information, media, and technology skills; and (3) Life and career skills. The fulfillment of these skills and abilities requires a learning process that provides the broadest possible possibility for students to creatively and innovatively learn to develop critical thinking skills to solve problems by utilizing various strategies and learning resources. With technology and information literacy skills, students are allowed to develop their potential, according to the learning objectives' competencies. The 21st-century teacher must be able to apply a learning model that uses a hybrid pattern (hybrid learning) because the learning process in the 21st century is conventionally face-to-face in class and online through the learning site. According to Garrison and Vaughan (2008), blended learning is more than improving lectures. It is a transformation in how we view teaching and learning. Blended learning is a systematic learning model that evaluates and combines face-to-face and online learning strengths to meet worthwhile educational objectives.

Several previous studies indicated that there was a positive impact of using blended learning on improving the quality of teaching and learning process, especially foreign languages. Behjat et al. (2012) discovered in a study in the Iranian tertiary education context that blending traditional classroom instruction and technology can help learners excel in their reading comprehension. In line with this, Ghazizadeh and Fatemipour (2017, p. 606) in their study on the effect of blended learning on EFL Learners' Reading Proficiency concluded that blended learning has a statistically significant positive effect on the reading proficiency of Iranian EFL learners. Blended learning can be adopted in English language classes, in order to facilitate the learning process especially that of the reading skill. Despite the benefits of blended learning in foreign language learning, as illustrated in the research results above, the use of blended learning in foreign language learning can be associated with several challenges and problems that need to be addressed by teachers. Alpala and Florez (2011) argue that although there is no "magic formula" for applying Blended

DOI 10.1201/9781003199267-15

Learning in general, the implementation of Blended Learning requires several considerations. One of them is the evaluation of the Blended Learning experience. Teachers should take into account not only their own views but other teachers' opinions and the feedback given by students in order to make some improvements in future implementations. This is important because blended learning is an approach to teaching that incorporates conventional face-to-face and online teaching in one setting.

COVID-19 pandemic changed the entire learning strategy. Teachers, lecturers, and students are forced to familiarize themselves with virtual learning. Student-centered learning that should be carried out by combining face-to-face and online learning must be done virtually. The evaluation of the effectiveness of the blended learning methods compared to fully online learning has not been widely available. The majority of the studies on blended learning above are aimed at describing the advantages of blended learning in increasing learning achievement that was integrated with the classroom. On the other hand, student perception as feedback could provide important information for the evaluation of foreign language learning strategies in order to improve the learning outcome. Therefore, this study aims to see how the correlation between students' perceptions of blended learning and their learning outcomes.

2 LITERATURE REVIEW

Blended learning is a learning method that supports the use of technology in conventional face-to-face learning. Graham (2006) suggested that blended learning systems mix face-to-face with computer-mediated learning interaction. As the current era progresses, teleconferencing becomes more common; thus, Friesen (2012) has proposed a new definition of 'face-to-face' (F2F) as 'co-present.' Friesen described blended learning as the variety of possibilities provided by integrating the Internet and digital media with existing types of a classroom that involve the physical co-presence of teachers. Technology plays a significant role in improving university learning experiences, improving student-to-teacher engagement, student-to-student contact, as well as time-to-work interaction, and helps address the demands of students with different learning styles (Kuh et al. 2011; Keshta & Harb 2013).

Previous research showed that students assume that providing a connection between the face-to-face and online learning environments in the integrated courses enables them to obtain information and input from different outlets, contribute to the course's subject matter, and develop trust in the practical implementation of gained knowledge. (Bliuc et al. 2011). The University of Central Florida applied blended learning and found that blended courses have higher completion rates than their equivalent face-to-face courses and entirely online courses (Dziuban et al. 2006). A study by Poon (2012) showed that blended learning creates a flexible learning atmosphere and facilitates better group work. Other studies have also shown an increase in interaction between students inside and outside the classroom because learning activities are not limited to the classroom (McLaughlin et al. 2015).

Researchers found a positive correlation between quality interactions on blended courses and student learning outcomes. They propose that this achievement is obtained because of the ability of students to conceptualize their learning material and be responsible for the entire process (Chou & Chou 2011; Moore & Gilmartin, 2010). Jachin and Usagawa (2017) also stated in his research that students in a blended class could achieve better grades than a traditional class and recommended that similar educational institutions in developing countries should implement blended learning. On the other hand, studies by Alonso et al. (2010) and Woltering et al. (2009) did not find significant differences in student test results between groups using the blended course and traditional methods. However, students generally have better motivation and satisfaction with blended learning.

In language learning, Albiladi and Alshareef (2019) presented that blended learning can help students to develop language skills, uplift the English learning environment, and encourage their motivation toward language learning. Blended learning can also improve students' language learning outcomes, as conveyed in research by Tomlinson and Whittaker (2013). In contrast, Tosun

Table 1. Respondent demographics by gender.

Gender	Total	Percentage
Male	9	28,13
Female	23	71,87
Total	32	100

(2015, p. 646) in his research found that blended learning strategy did not improve the students' vocabulary achievement. The result might be attributed to a short duration of the study, and a lack of self-discipline and motivation. Blended learning, when implemented properly, yields satisfactory results. Therefore, before being implemented, teachers must prepare the learning design properly to accommodate the various learning styles of students. The teaching style applied must also be made simple but attractive, so that students stay motivated and engaged during the learning process.

3 RESEARCH METHOD

This research aims to analyze the correlation between students' perception of blended learning and their learning outcome in the Aufbaustufe course. This research was held in the German Language Education Study Program of the Faculty of Teacher Training and Education at one of the public universities in Indonesia, the academic year of 2019/2020. We used the random sampling technique in this study, based on certain criteria according to research objectives. Participants of this study were third-semester students We used the random sampling technique in this study, based on certain criteria according to research objectives, and obtained 32 samples with details shown in Table 1.

It used a quantitative research method to analyze the correlation between the two variables. The data collected for students' perception was obtained through questionnaires. The questionnaire used in this study contains 25 items related to the availability of facilities, use of the internet, and e-learning literacy, and students' readiness to use the blended learning method in the learning process. Students' responses were scored based on a Likert scale with five points ranging from strongly disagree, disagree, neither/nor agree, agree, and strongly agree. Aufbaustufe is one of the subjects on intermediate-level German language skills taught at the German Language Education Study Program at the university mentioned above. The data collected for Aufbaustufe learning outcomes are gathered through some tests. All the collected data were analyzed by using the Pearson correlation technique with counting technique assisted by SPSS Application.

4 FINDINGS AND DISCUSSION

Based on the data we collected to examine students' perception on the implementation of blended learning, most of the participants (73%; 23 people) were responded with "strongly agree" with points 585/800, and six people (21%) were responded with "agree" with points 168/800, three people (6%) were responded with "disagree" with point 47/800. No participants responded with strongly disagree, and neither/nor agree (Figure 1).

The total score of the Perception score is 3627, with a mean of 117 ± 4.71, while the Aufbaustufe course test total score is 2662, with a mean of 85.87 ± 7.32. The data is then analyzed using the Pearson Product moment correlation formula, and obtained $r_{count} = 0.90$. When compared with $r_{table} = 0.361$, it turns out r_{count} $0.90 \geq r_{table} = 0.361$. The result means that there is a significant correlation between students' perceptions of the use of blended learning and Aufbaustufe learning outcomes (intermediate-level German language skills). Thus, the better the student's perception of the use of blended learning, the better Aufbaustufe learning outcomes achieved.

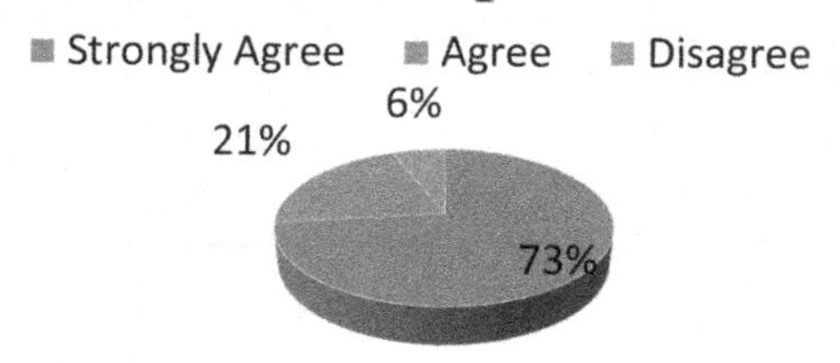

Figure 1. Students' perception toward blended learning model.

Table 2. The correlation between Students' Perception and Aufbaustufe Course Test.

Variables	Total Score	Mean	SD	Varian
Perception Score	3627	117.00	4.71	21.48
Aufbaustufe Test Score	2662	85.87	7.32	51.79
			Correlation	0.90

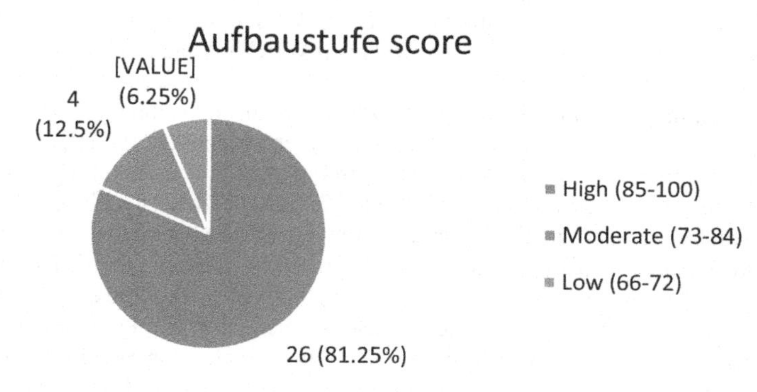

Figure 2. Aufbaustufe Test Score.

Our finding indicates that students' perceptions of blended learning contributed to the Aufbaustufe learning outcomes achieved. This finding means that if students have a positive perception of blended learning, the learning outcomes will be better. This can be seen in most students' perception scores, namely 73% who view blended learning as a learning model that can make them more creative and innovative, which is directly proportional to students' score for the Aufbaustufe course in the "high" category, namely 81.25% (Figure 2).

This analysis result indicates that students become more flexible with the learning process they are doing through blended learning. This is in line with the demands of the 21st-century learning paradigm conventionally face-to-face in class and online through various learning sites. Through a combined pattern of face-to-face meetings with online-based learning, using a computer or smartphone connected to an internet network can provide the widest possible opportunity for teachers and students to carry out learning activities and improve the quality of the learning process while doing other activities. This learning pattern fits perfectly with students' characteristics in this digital era who are very familiar with Information Technology.

Most students view blended learning as a learning model that can make them more creative and innovative. With blended learning, students become more flexible with the learning process they are doing. The limitations of time, place, methods of delivering materials, and communication style can be overcome through the use of technology. Students' perceptions of the blended learning we obtained in this study were very positive. From the data we obtained, six students answered that they agreed to the implementation of blended learning in the learning process and had not fully supported the use of this model. After further investigation, they think that direct communication is more important because they can discuss it easily. Meanwhile, only 6% or three students disagreed with the application of blended learning because it required more expenses, the availability of a more stable internet network, and had to be supported by an appropriate cell phone.

Wang et al. (2013) in their study found that students overall reacted positively to the use of the wiki, a web-based learning tool. A study by Miyazoe and Anderson (2010) also revealed students' positive perceptions of the blended learning model through the use of three different online writing tools in learning English as a foreign language. The study also revealed that there were positive effects obtained from using the three online writing tools on students' language learning. These studies are consistent with what we found in this study.

Another study presented that not all students responded positively to online learning. Students seemed to have critical reflections in face-to-face settings and elaborate the lesson much better than in online settings. Researchers say that students appear to be more worried about postings that are frequently deficient in deep thought (Bliuc et al. 2011). Garrison and Vaughan (2008) would argue that this would not happen if online debates are adequately organized and encouraged. Owston et al. (2013) found strong relationships between perceptions and learning outcomes. High-achiever students preferred blended learning to conventional learning styles. They also found integrated classes more comfortable, interactive, and it was easier to master the course topics than other conventional face-to-face classes. On the other side, low-achiever students found it hard to adapt their learning style to mixed learning. Thus, when scaling up blended learning institutions, it might be prudent to consider giving students the option of whether to enroll in blended or entirely face-to-face courses where possible, particularly in subjects that are challenging for students.

5 CONCLUSION

Our results reveal a significant correlation between students' perceptions of blended learning and their learning outcomes. The blended learning model has been widely recognized in the world of education around the world and has been applied to various subjects at various levels of education. Most of the research has also shown a positive correlation between blended learning and more satisfying learning outcomes. However, the use of blended learning also has drawbacks. Difficult access to technology and adequate internet networks is a challenge itself, especially in developing countries. Also, some students prefer face-to-face learning because of the difficulty of adapting to the use of technology and choosing direct communication with teachers and fellow students. This positive result of the study regarding the acceptance of virtual blended learning as a new learning strategy that was applied during the COVID-19 pandemic conditions indicated that educational institutions should provide more financing for the development of more integrated facilities and technology to apply the blended learning model. This study's weakness is the small number of samples and only examines the use of blended learning in one subject, so the results of this study cannot be broadly integrated. In future studies, it is hoped that more samples will be used and the need for comparative analysis of blended learning with conventional settings.

REFERENCES

Albiladi, W. S., & Alshareef, K. K. (2019). Blended learning in English teaching and learning: A review of the current literature. *Journal of Language Teaching and Research, 10*(2), 232–238.

Alonso, F., Manrique, D., Martínez, L., & Viñes, J. M. (2010). How blended learning reduces underachievement in higher education: An experience in teaching computer sciences. *IEEE Transactions on Education, 54*(3), 471–478.

Alpala, C. A. O., & Flórez, E. E. R. (2011). Blended learning in the teaching of English as a foreign language: An educational challenge. *How, 18*(1), 154–168.

Behjat, F., Yamini, M., & Bagheri, M. S. (2012). Blended learning: A ubiquitous learning environment for reading comprehension. *International Journal of English Linguistics, 2*(1), 97.
Bliuc, A. M., Ellis, R. A., Goodyear, P., & Piggott, L. (2011). A blended learning approach to teaching foreign policy: Student experiences of learning through face-to-face and online discussion and their relationship to academic performance. *Computers & Education, 56*(3), 856–864.
Chou, A. Y., & Chou, D. C. (2011). Course management systems and blended learning: An innovative learning approach. *Decision Sciences Journal of Innovative Education, 9*(3), 463–484.
Dziuban, C., Hartman, J., Juge, F., Moskal, P., & Sorg, S. (2006). Blended learning enters the mainstream. In C. J. Bonk & C. R. Graham (Eds.), *The Handbook of Blended Learning: Global Perspectives, Local Designs* (pp.195–206). San Francisco, CA: Pfeiffer Publishing
Friesen, N. (2012). Report: Defining Blended Learning.
Garrison, D. R., & Vaughan, N. D. (2008). *Blended learning in higher education: Framework, principles, and guidelines*. New Jersey: John Wiley & Sons.
Ghazizadeh, T., & Fatemipour, H. (2017). The effect of blended learning on EFL learners' reading proficiency. *Journal of Language Teaching and Research, 8*(3), 606–614.
Graham, C. R. (2006). Blended Learning Systems: Definition, Current Trends, and Future Directions. In C. J. Bonk & C. R. Graham (Eds.), *The Handbook of Blended Learning: Global Perspectives, Local Designs* (pp.3–21). San Francisco, CA: Pfeiffer Publishing
Jachin, N., & Usagawa, T. (2017). Potential impact of blended learning on teacher education in Mongolia. *Creative Education, 8*(09), 1481.
Keshta, A. S., & Harb, I. I. (2013). The effectiveness of a blended learning program on developing Palestinian tenth graders' English writing skills. *Education Journal, 2*(6), 208–221.
Kuh, G. D., Kinzie, J., Schuh, J. H., & Whitt, E. J. (2011). Student success in college: Creating conditions that matter. New Jersey: John Wiley & Sons.
McLaughlin, J. E., Gharkholonarehe, N., Khanova, J., Deyo, Z. M., & Rodgers, J. E. (2015). The impact of blended learning on student performance in a cardiovascular pharmacotherapy course. *American Journal of Pharmaceutical Education, 79*(2).
Miyazoe, T., & Anderson, T. (2010). Learning outcomes and students' perceptions of online writing: Simultaneous implementation of a forum, blog, and wiki in an EFL blended learning setting. *System, 38*(2), 185–199.
Moore, N., & Gilmartin, M. (2010). Teaching for better learning: A blended learning pilot project with first-year geography undergraduates. *Journal of Geography in Higher Education, 34*(3), 327–344.
Owston, R., York, D., & Murtha, S. (2013). Student perceptions and achievement in a university blended learning strategic initiative. *The Internet and Higher Education, 18*, 38–46.
Poon, J. (2012). Use of blended learning to enhance the student learning experience and engagement in property education. *Property management, 30(2)*, 129–156.
Tomlinson, B., & Whittaker, C. (2013). *Blended learning in English language teaching.* London: British Council.
Tosun, S. (2015). The effects of blended learning on EFL students' vocabulary enhancement. *Procedia-Social and Behavioral Sciences, 199*(1), 641–647.
Trilling, B., & Fadel, C. (2012). 21st century skills: Learning for life in our times. New Jersey: John Wiley & Sons.
Wang, J., Zou, B., Wang, D., & Xing, M. (2013). Students' perception of a wiki platform and the impact of wiki engagement on intercultural communication. *System, 41*(2), 245–256.
Woltering, V., Herrler, A., Spitzer, K., & Spreckelsen, C. (2009). Blended learning positively affects students' satisfaction and the role of the tutor in the problem-based learning process: results of a mixed-method evaluation. *Advances in Health Sciences Education, 14*(5), 725–738.

Post Pandemic L2 Pedagogy – Adi Putra & Arifah Drajati (Eds)

A comparative study of Jigsaw and student team achievement division techniques in writing narrative text

Ika Purnama Sari
STIKOM Tunas Bangsa Pematangsiantar, Indonesia

Susiani
AMIK Tunas Bangsa Pematangsiantar, Indonesia

ABSTRACT: The study aims to introduce how a lesson built on the improvement of different intelligence levels through cooperative learning can be implemented with Jigsaw and STAD techniques. This study design is a quantitative approach with an experimental design. This study uses Pretest and Posttest scores. The participants consist of 44 students. The data was collected from students' tests. T- test was applied to calculate the data and to test the hypothesis. The study finding indicated that Significant value (2 – tailed) is 0.023 which is <0.05, and T_{value} is 2.365 > T_{Table} 1.724 which means that the Null Hypothesis (H_o) is rejected at the Level of Significant 0.05. The mean of Jigsaw technique is 82.23 and the mean of STAD technique is 77.18. This study is done for planning and organizing teamwork and for supporting learning.

Keywords: Cooperative Learning, Jigsaw, Students Teams – Achievement (STAD), Writing Narrative Text.

1 INTRODUCTION

Cooperative learning improves students' mindset with their friends in a group, (Chen 2018). There are a great number of cooperative learning techniques available. Additionally, (Arends 2012) gives the explanation of cooperative learning techniques that consist of STAD, Jigsaw, Group Investigation, The Structural Approach, Think Pair share, and Numbered heads Together. The Jigsaw technique was developed by Elliot Aronson (Aronson 1978) and the Student Teams Achievement Division (STAD) technique was developed by Robert Slavin (Slavin 1994). Here the researcher focuses on Jigsaw and STAD teaching techniques. Even though the Jigsaw and STAD were part of Cooperative learning. The researcher focused on grouping them. Based on the previous research before, there is a positive effects of cooperative learning on student achievement (Roseth et al. 2008); (Stevens & Slavin 1995); (Mattingly & VanSickle 1991).

Writing ability of a second semester student in one University in Indonesia is still low. It might be caused by a number of factors like study habit, self-confidence, concentration, teaching and learning facilities, teachers or the learning environment. The study aims to introduce how a lesson built on the improvement of different intelligence levels through Cooperative Learning can be implemented with Jigsaw and STAD techniques. After this study, we know what techniques of Cooperative Learning are more significant for students' achievement in writing narrative by using Jigsaw or STAD technique.

2 LITERATURE REVIEW

Text is a unit of meaning for its context. Mostly people think that it is only in the form of a written work, but the text can be spoken or written. A text is any stretch of language, (Feez & Joyce 1998).

DOI 10.1201/9781003199267-16

We are creating and constructing a text when we use language to write. We are interpreting texts when we read. Then, we are also creating and interpreting texts when we talk and listen. Although sometimes there are stories which are built based on real experiences or events, Narrative texts are imaginary stories with the aim to entertain (Harahap et al. 2019). Narrative is also a genre. It can easily accommodate one or more of the other genres and still remain dominant, (Knapp & Watkins 2005).

Cooperative learning refers to work in small groups to help each other learn (Slavin 2015). Cooperative Learning is a method of learning by forming groups. Cooperative Learning showed teachers organizing students into small groups then work together to help one another learn academic content (Tran et al. 2019). In this group students who have different understandings are encouraged to teach each other. Cooperative Learning is an approach that makes maximum use of cooperative activities (Richards & Rodgers 2014). Cooperative Learning is more than group work; it is a group work designed to nurture strong social interdependence amongst students, (Johnson et al. 2007).

Jigsaw was developed by Elliot Aronson (Aronson 1978). Jigsaw technique is an alternative teaching method. The students are taught to work in smaller interdependent groups (Aronson 1978). The steps are as follows: students are grouped with members of ±4 persons, each person on the team is given different materials and tasks, members of different teams with the same assignment form a new group (expert group), after expert group discuss, each member Go back to the original group and explain to the group members about the sub-section they are mastering, each expert team presents the results of discussion, discussion, closing. In Jigsaw teaching technique, the students are set up in teams, each team member responsible for mastering part of the learning material, (Huda 2011). Jigsaw has more complex group structures (Chang & Benson 2020).

Student Teams Achievement Division (STAD) was developed by Robert Slavin (Slavin 1994). The STAD model contains cooperative learning steps of the delivery of goals and motivation, group division, percentage of teachers, team learning activities (teamwork), evaluation, team achievement awards. Student Teams Achievement Division (STAD) is one of a set of instructional techniques collectively known as Student team Learning. Student Team Learning techniques are not one-time activities designed to liven up the classroom from time to time, but can be used as permanent means of organizing the classroom to effectively teach a wide variety of subjects, (Pedersen & Digby 2014). Student Teams Achievement Divisions (STAD) has been influential in bringing positive effects in multiple grades and students, (Alijanian 2012).

3 RESERACH METHOD

3.1 *Study design*

This study was conducted by using Experimental Design. Experiment is the process of examining the truth of statistical hypothesis, (Kothari 2004).

The participants of this research were the second semester students in one University in Indonesia are 342 students. The researcher used two classes. One class was taught by Jigsaw and the other by STAD technique.

Table 1. Table study design.

Group	Pre Test	Treatment	Post Test
Group 1	Y_1	T1, T2, T3… of Jigsaw Technique	Y_2
Group 2	Y_1	T1, T2, T3… of STAD Technique	Y_2

3.2 *The instrument*

The instrument used by the writer in this study is a test. Writing a narrative text was used as the instrument of the study Generic structure and Lexicogrammatical Features.

Table 2. Assessment aspect of writing narrative essay.

No	Aspects Assessed	Score
1	Social Function (Content)	30
2	Generic Structure (Organization)	20
	a. Orientation	
	b. Complication	
	c. Resolution	
3	Vocabulary	20
4	Grammatical Features	25
	a. Action Verbs	
	b. Relational Verbs	
	c. Simple Past Tense	
5	Mechanic (Spelling & Punctuation)	5
	Total	100

3.3 *Technique of data collection*

The technique of data collection in this study used quantitative data. The test was distributed through pre-test and post-test. The researcher applied for a writing test. To administer the writing test, the researcher uses an analytic score in order to be more reliable in scoring students' writing. The score consists of: Social Function (Content) + Social Function (Organization) + Vocabulary + Grammatical Features + Spelling & Punctuation = Total score.

Table 3. Total score.

Categorization	**Score**
Very Weak	Score 10–30
Weak	Score 31–55
Enough	Score 56–75
Good	Score 76–85
Excellent	Score 86–100

3.4 *Technique of data analysis*

In analysis the data, the study was tested: description analysis for describing the study data including Mean, Median, Mode, Variance and Standard Deviation. The data is presented in the table of frequency distribution and Histogram by using the SPSS program and Inferential analysis for measuring the Hypothesis which is done by ANOVA.

4 FINDINGS AND DISCUSSION

4.1 *Result of pre-test and post test of Jigsaw and STAD techniques*

In the pre-test of Jigsaw technique, the mean is 46.86, the median is 47.00, the variance is 15.838 and the Standard Deviation is 3.980. In the post-test of the Jigsaw technique, the mean is 82.23, the median is 82.50, the variance is 37.136 and the Standard Deviation is 6.094. In the pre-test of

STAD technique, the mean is 45.95, the median is 44.00, the variance is 65.950 and the Standard Deviation is 8.121. In the post-test of the STAD technique, the mean is 77.18, the median is 77.00, the variance is 63.013 and the Standard Deviation is 7.938.

4.2 *Testing of normality*

The Normality Test aims at showing that the sample data of the study is normality distributed. The normality test applied in this study Kolmogorov-Smirnov and Shapiro–Wilk. Here the researcher uses a sample of less than 50 students, so the researcher focuses on Shapiro–Wilk, because the sample is less than 50 students. The results of the Normality Test of the students' show that the Sig. value of Jigsaw Teaching Technique is 0.693 which is >0.05, the Sig. value of the STAD Teaching Technique is 0.132 which is >0.05. It can be seen in Table 4 as below.

Table 4. Testing of normality.

		Kolmogorov-Smirnov[a]			Shapiro-Wilk		
	Kelas	Statistic	df	Sig.	Statistic	df	Sig.
Hasil	Jigsaw	.167	22	.113	.969	22	.693
	STAD	.168	22	.107	.931	22	.132

a. Lilliefors Significance Correction

Based on the result of calculation described in Table 8 above and the criteria of Normality Test, it is concluded that all data in this study had normal distribution.

4.3 *Testing of homogeneity*

The Homogeneity Test aims at investigating whether variance of the data is homogeneous. The Homogeneity Test of variance was calculated by using Levene test by using SPSS 19.00 program for learning model and students' personality and interaction groups. The results of the Homogeneity Test of the students' show that Sig. value based on mean is 0.396 which is >0.05, the Sig. value based on Median is 0.404 > 0.05, the Sig. value based on Median and with an adjusted df is 0.404 > 0.05 and Sig. the value based on trimmed means 0.388 which is >0.05 This can be seen in Table 5.

Table 5. Testing of homogeneity.

		Levene Statistic	df1	df2	Sig.
Students' Achievement	Based on Mean	.735	1	42	.396
	Based on Median	.711	1	42	.404
	Based on Median and with adjusted df	.711	1	39.005	.404
	Based on trimmed mean	.761	1	42	.388

Based on the computation of the Homogeneity Test above, Thus variance is Homogeneous.

4.4 *Testing hypothesis*

The researcher then calculates the data T-test the hypothesis. This was the last calculation and this was a crucial calculation to answer the problem formulation of difference between students' ability in writing Narrative text were given Jigsaw technique and STAD technique. The result of t–test on post-test of both classes is described in Table 6 and Table 7 below.

Table 6. Independent sample T test.

Independent Samples Test

		Levene's Test for Equality of Variances		t-test for Equality of Means						
									95% Confidence Interval of the Difference	
		F	Sig.	t	df	Sig. (2-tailed)	Mean Difference	Std. Error Difference	Lower	Upper
Student's achievement	Equal variances assumed	.735	.396	2.365	42	.023	5.045	2.134	.740	9.351
	Equal variances not assumed			2.365	39.372	.023	5.045	2.134	.731	9.360

Table 7. Group statistic.

Group Statistics

	Kelas	N	Mean	Std. Deviation	Std. Error Mean
Student's achievement	Post Test Jigsaw	22	82.23	6.094	1.299
	Post Test STAD	22	77.18	7.938	1.692

Based on the result calculation described in Table 11 above, it is concluded that the Significant value (2 – tailed) is 0.023 which is <0.05, and T_{value} is 2.365 > T_{Table} 1.724. Thus, Null Hypothesis (H_o) is rejected at the Level of Significance 0.05.

Based on Table 7 above, there is a significant difference between students' ability in writing narrative text. The mean of the Jigsaw technique is 82.23 and the mean of STAD technique is 77.18, it is concluded that the Jigsaw technique is more effective to enhance students' ability to write narrative text

The result of this study of Cooperative Learning was consistent with the generally superior academic achievement effects reported in the cooperative learning studies reviewed by (Roseth et al. 2008); (Stevens & Slavin 1995) and (Mattingly & VanSickle 1991). In (Roseth et al. 2008)'s result showed that the true effect was larger than the substantial large effect observed. In (Stevens & Slavin 1995)'s results show that Cooperative Learning classes had significantly higher achievement. In (Mattingly & VanSickle 1991)'s result generally superior academic achievement effects. The result of this research showed that there is interaction between Jigsaw and STAD Teaching Technique on Students' Narrative Writing Achievement. There are some Teaching Techniques in Cooperative Learning Method, but in this research the researcher used 2 Teaching Techniques. Based on the explanation above the research assumes that Jigsaw Teaching Technique is more significant to affect the students' ability in Writing Narrative Text than STAD Teaching Technique. We concluded that Cooperative Learning is effective to enhance students' ability.

5 CONCLUSION

Based on the data analysis of the data post-test, there is a significant difference between students' achievement in writing narrative by using group Jigsaw and STAD techniques. The average score of the class which applied STAD technique is lower than the average score which applied Jigsaw technique. By the difference grouping we can see the different results. Here, we focus on the technique of grouping Jigsaw and STAD. From the above findings it can be concluded that Jigsaw technique is more effective to enhance students' ability in writing narrative text. In line with the conclusion drawn, the suggestions below are: English teachers are suggested to apply Cooperative Learning to enhance the result of students' achievement in writing Narrative Text, then the teachers are expected to be creative in motivating their students to write, and the studier who wants to conduct further to make improvement in this study by developing a study focus to other types of relevant learning techniques to study so that it can be used as an alternative in improving students' achievement in writing narrative text.

REFERENCES

Alijanian, E. (2012). The effect of student teams achievement division technique on English achievement of Iranian EFL learners. *Theory & Practice in Language Studies, 2*(9).

Arends, R. I. (2012). *Learning to teach . New York: Mcgraw-Hill Companies*. Inc.

Aronson, E. (1978). *The jigsaw classroom.* Sage.

Chang, W.-L., & Benson, V. (2020). Jigsaw teaching method for collaboration on cloud platforms. *Innovations in Education and Teaching International*, 1–13.

Chen, Y. (2018). Perceptions of EFL college students toward collaborative learning. *English Language Teaching, 11*(2), 1–4.

Feez, S., & Joyce, H. D. S. (1998). Text-based syllabus design. *National centre for English language teaching and research*, Macquarie University.

Johnson, D. W., Johnson, R. T., & Smith, K. (2007). The state of cooperative learning in postsecondary and professional settings. *Educational Psychology Review, 19*(1), 15–29.

Knapp, P., & Watkins, M. (2005). *Genre, text, grammar: Technologies for teaching and assessing writing.* UNSW Press.

Kothari, C. R. (2004). Research methodology: Methods and techniques. *New Age International.*

Mattingly, R. M., & VanSickle, R. L. (1991). Cooperative learning and achievement in social studies: jigsaw II.

Pedersen, J. E., & Digby, A. D. (2014). Secondary schools and cooperative learning: Theories, models, and strategies. *Routledge.*

Richards, J. C., & Rodgers, T. S. (2014). Approaches and methods in language teaching. *Cambridge university press.*

Roseth, C. J., Johnson, D. W., & Johnson, R. T. (2008). Promoting early adolescents' achievement and peer relationships: The effects of cooperative, competitive, and individualistic goal structures. *Psychological Bulletin, 134*(2), 223.

Slavin, R. E. (1994). Student teams-achievement divisions. *Handbook of Cooperative Learning Methods*, 3–19.

Slavin, R. E. (2015). Cooperative learning in elementary schools. *Education 3-13, 43*(1), 5–14.

Stevens, R. J., & Slavin, R. E. (1995). The cooperative elementary school: Effects on students' achievement, attitudes, and social relations. *American Educational Research Journal, 32*(2), 321–351.

Tran, V. D., Nguyen, T. M. L., Van De, N., Soryaly, C., & Doan, M. N. (2019). Does cooperative learning may enhance the use of students' learning strategies?. *International Journal of Higher Education, 8*(4), 79–88.

Post Pandemic L2 Pedagogy – Adi Putra & Arifah Drajati (Eds)

Developing students' critical thinking skills through culture-based instructional materials in EFL reading and writing courses

Haerazi, Zukhairatunniswah Prayati, Lalu Ari Irawan & Rully May Vikasari
Universitas Pendidikan Mandalika, Indonesia

ABSTRACT: This study is aimed at investigating the culture-based instructional materials (C-BIM) to trigger students' reading and writing skills viewed from critical thinking skills. This study is categorized as mixed-method using the explanatory sequential design. The data consist of quantitative and qualitative data. The instruments employ reading and writing tests, questionnaires, and interview sheets. The data of reading-writing achievement are computed using descriptive and inferential statistics. Thus, the data of students' and teachers' intercultural awareness and perception on culture-based materials are analyzed using qualitative steps that are analysis, reduction, coding, interpretation, and taking a conclusion. Based on the result of analysis, this study shows that the culture-based instructional materials have a significant effect on students' reading and writing skills. This study offers implications and recommendations for further research and practice in marrying culture-based materials in reading and writing courses.

Keywords: Critical Thinking Skills; Culture-Based Materials; L2 Reading, L2 Writing

1 INTRODUCTION

Students at the middle schools in Indonesia are required to have higher-order thinking skills as stated in the Indonesian National curriculum. One of those is a critical thinking skill. Becoming a critical thinking learner is commanded in the National curriculum (K13.) To attain this goal, English teachers are demanded to find out appropriate ways to teach. Critical thinking skills require students to analyze and evaluate learning activities and justify conclusions (Manshaee et al. 2014); Thompson & Patterson 2019). In other words, critical thinking is a habit of mind to analyze and gauge evidence and confirm conclusions. For English learners in middle schools, they are trained to think critically through how they organize an essay, analyze language features, and monitor their learning activities.

Previous studies showed that critical thinking skills have an effect on ELT students' language skills and achievement. Liu and Stapleton (2018) declared that critical thinking skills help students to think effectively how they write essay organization, lexical items, and meta-discourses. Yuan and Stapleton (2020) argued that ELT teachers have a positive perception of critical thinking to help them monitor their learning achievement; it means that critical thinking skills help students and teachers develop their creativity in learning and teaching. It is also in line with Espey (2018) who depicted that critical thinking skills such as the analyzing and problem-solving activities improve students' academic achievement. In addition, Low et al. (2017) noted that it could develop student teachers' professional and pedagogical knowledge. It indicated that critical thinking is essential in ELT courses.

DOI 10.1201/9781003199267-17

2 LITERATURE REVIEW

Given the importance of critical thinking skills, this study aimed to enhance students' critical thinking skills to help them improve their reading and writing skills. Reading and writing skills are two of the language skills necessary to improve (Haerazi et al. 2020). Reading and writing skills interplayed with each other. The more you read, the more insights you have to write. Unfortunately, most students have a negative attitude towards reading because reading activities sometimes are boring (Wiles et al. 2016). Compared to reading skills, students need hard work to acquire writing skills. Writing requires more knowledge of linguistics, cognitive use, and cultural dimensions.

Insertion of culture-based teaching materials in the teaching of reading and writing is a way to bring students to think critically toward their own cultures. The cultural topics in this study include local traditional arts, traditional games, traditional rites, traditional technologies, and local literacies. According to Dinh and Sharifian (2017), the insertion of local cultures into EFL textbooks can raise students' cultural practice awareness and concepts. Also, it enables them to discuss and talk about cultural conceptualizations, to promote intercultural contacts, and to develop their cultures and cultural identities. It is supported by McConachy and Hata (2013); and McConachy 2018) who state that culture-based EFL materials encourage students to reflect their own cultures into their cultural practices, whether orally or in written

Given the importance of culture-based instructional materials, this study is aimed at developing the culture-based materials for teaching reading and writing at middle schools. The novelty of this study lies on the use of critical thinking skills by incorporating culture-based instructional materials to enhance students' reading and writing skills. Critical thinking is considered as the attribute variable which affects students' reading and writing achievement (McConachy 2018). This study initially started from designing the reading and writing tasks and materials to trigger students' critical thinking skills to facilitate students in reading activities and writing exercises.

3 RESEARCH METHOD

The current study was conducted at junior high schools. 350 students were involved in this research. There were 200 females and 150 males who were involved in the filling questionnaire items. Moreover, some of them also were involved in the interview sessions. The researcher decided the middle schools as a place for the field testing. The students there consisted of two classes in which each class comprised 29 students. One class was subjected to an experimental class using the C-BIM model and the second class was treated as the control class using the existing teaching materials.

It was classified as a mixed-method using the explanatory sequential design. This design was used to elaborate and explain the research data based on the research problems (Creswell & Creswell 2018). The data consisted of quantitative and qualitative data. The process of collecting data was started from quantitative data and then qualitative data. The instruments of collecting the quantitative data used tests and questionnaires. The instruments were stated as valid and reliable instruments. The quantitative data were computed using descriptive and inferential statistical analysis. The descriptive statistical analysis was done to find out mean, median, mode, and standard deviation of students' reading achievement, writing scores, and critical thinking skills, both experimental and control classes. Meanwhile, the inferential statistical analysis was aimed at finding out normality distribution, homogeneity of variance, and multifactor analysis of variance.

4 FINDINGS AND DISCUSSION

The place of culture in English language teaching becomes complex issues. Various ways to teach are done by English teachers around the world, including incorporating local cultures and target cultures in their textbooks. This study aimed to investigate; (1) the effectiveness of the culture-based instructional materials (C-BIM) to trigger students' reading and writing skills viewed from critical thinking skills at middle schools, Central Lombok, West Nusa Tenggara, Indonesia; (2) the teachers' intercultural awareness and the students' and teachers' perception on incorporating culture-based instructional materials to enhance students' reading and writing skills in relation to critical thinking skills at the middle schools. The issues of the effectiveness of the C-BIM model were elaborated quantitatively, while the issues of the English teachers' intercultural awareness and the students' and teachers' perception of the C-BIM model were exposed qualitatively.

4.1 *The effectiveness of the C-BIM model*

Some studies indicated that culture-based teaching materials can help students improve their language skills such as speaking skills (Fernández-Agüero & Chancay-Cedeño 2019; Aprianoto & Haerazi 2019), reading comprehension and writing skills (Haerazi & Irawan 2020). Also, it can develop cultural and intercultural awareness (McConachy 2018), multicultural insight (Uzum et al. 2018), and linguistic awareness (Krulatz et al. 2018). In this current study, researchers start from the development process to find out the effectiveness of the C-BIM model. The teaching materials of reading and writing skills are developed in line with the current curriculum at the middle schools. The used teaching materials are inserted cultures, local cultures, and target cultures. Based on the initial need analysis, the selected cultures include traditional arts, traditional games, traditional rites, traditional technologies, and local literacies.

The researchers attained the students' reading and writing achievement using reading and writing tests, respectively. To see the students' critical thinking skills, researchers employed a test along with reading and writing tests. To control the effectiveness of the C-BIM model, this study treated students in the experimental class using the C-BIM model, while students in the control class applied the existing learning model, non-cultural materials. The results of the two treatments are computed using statistical analysis. Table 1 illustrates the students' average scores in the experimental and control classes. Thus, the multifactor analysis of variance (ANOVA) was carried out to know the proposed hypotheses.

The students' achievement in reading, writing, and critical thinking skills in the experimental class was greater than the control class. It means that the culture-based materials are more effective than non-culture-based teaching materials in enhancing students' reading and writing skills viewed from their critical thinking. It was supported by Shin et al. (2011) who claimed that culture-based materials or culture-based textbooks in EFL class could develop students' language skills, including reading and writing skills. In reading activities, students are exposed to various cultural texts. They are asked to read sentences and analyze the structure of the sentences. In pairs, they discuss the ideas from the text, notice the pattern of sentences and paragraphs, and communicate

Table 1. Results of students' achievement in the experimental and control classes.

Skills	Groups and Level	Mean
Reading Skills	C-BIM Model	72.183
	Non-Culture	61.596
Writing Skills	C-BIM Model	75.593
	Non-Culture	65.553
Critical Thinking Level in Reading Classes	High	75.834
	Medium	65.333
	Low	59.500
Critical Thinking Level in Writing Classes	High	78.205
	Medium	67.597
	Low	65.917

their comprehension. They are involved in the process of intercultural communication in reading activities.

During the reading activities, students listed some vocabulary that they do not understand from the texts. They are allowed to open an English-Indonesia dictionary. In groups, students compare the cultural dimension they get from the text with their own cultures. In doing so, the teacher guides students with lists of cultural values such as responsibility, discipline, custom, belief, and tolerance. Thus, they are involved in a discussion group. For instance, they interpret the values of traditional music such as *Gendang Beleq* for the Sasak community. Some argue that this tradition indicates the *Sasaq* people have an act of courage and a beautiful attitude for anyone. The other student states it illustrates *Sasaq* people should have the courage to fight. This comprehension is then noticed in sentences. Building students' prior knowledge is crucial things like a piece of information to conduct sentences later on in writing activities. It is in accordance with Haerazi et al. (2020), who found building knowledge of the field before writing is necessary because students are involved in developing their cognition of what they are going to write.

During writing exercises, students are asked to create sentences based on their reading comprehension. In doing so, they are completed with a graphic organizer. Using graphic organizers is important for students at the level of middle schools. According to Juniarti et al. (2017), the graphic organizer is useful in writing exercises for students to compose their ideas into a complete paragraph. The text type going to create in this current study is a complete descriptive text. The cultural topics read by students are enclosed with questions that bring students to think critically before writing. In the introductory paragraph of the descriptive text, students are demanded to create drafts in line with their reading comprehension in the forms of sentences.

To help students write this introductory part, the teacher gives students clues. The clues are designed in jumbled sentences. In pairs, students arrange their sentences into a paragraph. In this learning, students should pay attention to generic structures, grammars, and patterns of sentences, whether active or passive sentences. At the end of this phase, students are able to generate the complete text. In these activities, critical thinking activities seem helpful for students to elaborate, generate, evaluate, and conclude the generic structures of texts and reading comprehension. It is in keeping with Boyle et al. (2019) who argue that critical thinking skills can help them to generate qualified writing products.

To find out the interaction and significant difference between teaching materials and critical thinking skills in reading classes and writing classes, researchers employed the multifactor analysis of variance (ANOVA). Based on ANOVA analysis, the results informed that there is a significant difference between the C-BIM and non-cultural materials in enhancing students' reading comprehension and writing skills viewed from critical thinking skills. Also, the critical thinking skills determined the success of students' reading and writing skills at the eight-grade students. It could be stated that it had a significant impact on reading and writing skills. It was supported by some previous studies that informed that critical thinking had a big contribution to develop students' language skills such as reading skills (Kubashi & Fahmie 2020). Furthermore, the acquisition of critical thinking skills in reading and writing classes is influenced by the teaching materials and instructional methods or strategies the English teachers applied in classes.

4.2 *The teachers' intercultural awareness and the students' and teachers' perception on the C-BIM model*

English teachers are able to promote students' reading comprehension and writing skills using culture-based instructional materials in this study inasmuch as they have great intercultural awareness. Also, they and the students completely have perceived the use of culture-based instructional materials (C-BIM) model employed inasmuch as it can enhance students' English language proficiency, motivation, engagement, and self-efficacy. This study demonstrates that the teachers' intercultural awareness consists of dimensions of knowledge, attitude, interpretation skill, and cultural practice.

That the abilities of teachers perceived the intercultural dimensions in applying the C-BIM model was able to promote students' linguistic awareness inasmuch as they internalize and imitate the target language they learn. Providing various cultural texts can help students enhance their reading comprehension. Besides, intercultural activities explored in the class can help students to have skills of discovery and interaction. It is in line with Fernandez-Aguero and Chancay-Cedeno (2019) who argue that language learners can conduct a communication that represents the balance between how they perceive their cultures and how they are able to relate the local events with others from a different culture.

In this study, intercultural awareness plays a crucial role in applying the culture-based instructional materials (C-BIM) model in enhancing students' reading and writing skills. Another aspect as a moderator variable in this current study is critical thinking ability. This result was underpinned by the students' and teachers' positive perception of the C-BIM model inasmuch as it can help them to promote their language skills, engagement, motivation, and self-efficacy.

The culture-based instructional materials help students and teachers to arrange and administer their learning process. Teachers feel it is easy to improve students' reading comprehension and writing skills as well. It accords with some previous studies cited in this study on the use of culture and intercultural language learning (Raigón-Rodríguez 2018) The teacher in the class guides students to analyze any single word, phrase, and sentence in pre-reading activities. In doing so, students are completed with a vocabulary list. Dealing with the engagement, the C-BIM model is highly engaging students in learning activities. Some relevant studies inform that reading and writing can engage students in dynamic learning (Chen 2018; and Haerazi, Permadi, & Hidayatullah, 2020).

In terms of motivation, intercultural language learning motivated students to know and internalize their own cultures. As students are asked to identify and discuss their own local cultures, they are able to accomplish their reading tasks and create some drafts that are used to complete their writing later on. It indicates they feel very motivating in finishing the given tasks. It is in line with Song (2019) who reports that the cultural contents not only motivate students to associate their own knowledge to the target cultures but also help them accommodate cultural knowledge to concrete communication and imitate how English speakers express their ideas in written forms (Munandar & Ulwiyah 2012).

Students also have self-efficacy in the learning process under culture-based instruction. Due to good self-efficacy, students make progress in their reading and writing learning. In the learning process, students are asked to accomplish reading activities in a certain time. Surprisingly, they are able to finalize their tasks in accordance with the enacted time. Besides, culture-based instruction allows students to promote collaborative learning activities. It escalates students' mastery of fluency and readiness. In groups, students can acquire new vocabulary from their partners and also from the teachers' feedback. Because of these activities, the teacher feels easy to build students' self-confidence and self-efficacy to complete their learning assignments inside and outside classes. It is in accordance with some previous studies reporting that students' partnership and teacher's participation can improve students' self-efficacy and motivation (Burić & Kim 2020).

5 CONCLUSION

This study has provided empirical evidence on meaningful insertion and implementation using culture-based instructional materials in improving reading and writing skills. Another aspect of this study is critical thinking skills affecting the students' reading and writing skills. The findings show that culture-based materials positively contribute to processes of pre-reading, whilst reading, and post-reading activities. Also, students can realize their reading comprehension into a complete descriptive paragraph. Through cultural instruction, students can acquire and enhance their linguistic, cognitive, and sociocultural competences. It promotes students to develop reading and writing skills, motivation, engagement, and self-efficacy. For future studies, other researchers need to investigate the impact of cultural materials on students' speaking and listening skills viewed from socio cultural competences. In short, the current study is the catalyst for further practices and studies in developing culture-based instructional materials in the realm of ELT.

REFERENCES

Aprianoto, & Haerazi. (2019). Development and assessment of an intercultural-based instrument model in the teaching of speaking skills. *Universal Journal of Educational Research, 7*(12), 2796–2805.

Boyle, J., Ramsay, S., & Struan, A. (2019). The Academic Writing Skills Programme: A model for technology-enhanced, blended delivery of an academic writing programme The Academic Writing Skills Programme: A model for technology-. *Journal of University Teaching and Learning Practice, 16*(4).

Burić, I., & Kim, L. E. (2020). Teacher self-efficacy, instructional quality, and student motivational beliefs: An analysis using multilevel structural equation modeling. *Learning and Instruction, 66*(December 2019), 101302.

Chen, I. C. (2018). Incorporating task-based learning in an extensive reading programme. *ELT Journal, 72*(4), 404–414.

Creswell, J. W., & Creswell, J. D. (2018). Research Design: Qualitative, Quantitative, and Mixed Methods Approaches (Fifth Edition). In *Journal of Chemical Information and Modeling* (Vol. 53, Issue 9). Los Angeles: Sage Publication.

Dinh, T. N., & Sharifian, F. (2017). Vietnamese cultural conceptualisations in the locally developed english textbook: A case study of 'lunar new year'/'tet.' *Asian Englishes, 19*(2), 148–159.

Espey, M. (2018). Enhancing critical thinking using team-based learning. *Higher Education Research and Development, 37*(1), 15–29.

Fernández-Agüero, M., & Chancay-Cedeño, C. (2019). Interculturality in the Language Class – Teachers' Intercultural Practices in Ecuador. *RELC Journal, 50*(1), 164–178.

Haerazi, & Irawan, L. A. (2020). The effectiveness of ECOLA technique to improve reading comprehension in relation to motivation and self-efficacy. *International Journal of Emerging Technologies in Learning, 15*(1), 61–76.

Haerazi, H., Irawan, L. A., Suadiyatno, T., & Hidayatullah, H. (2020). Triggering preservice teachers' writing skills through a genre-based instructional model viewed from creativity. *International Journal of Evaluation and Research in Education, 9*(1), 234–244.

Haerazi, H., Utama, I. M. P., & Hidayatullah, H. (2020). Mobile applications to improve English writing skills viewed from critical thinking ability for pre-service teachers. *International Journal of Interactive Mobile Technologies (IJIM), 14*(07), 58.

Juniarti, K., Sofyan, D., & Kasmaini. (2017). The effect of using a graphic organizer to students' writing ability. *Journal of English Education and Teaching, 1*(1), 48–57.

Kubashi, H., & Fahmie, G. (2020). An investigation of the impact of critical thinking skills instruction on the Iraqi EFL learners' reading comprehension proficiency. *International Journal of Innovation, Creativity and Change, 11*(1), 363–372.

Krulatz, A., Steen-Olsen, T., & Torgersen, E. (2018). Towards critical cultural and linguistic awareness in language classrooms in Norway: Fostering respect for diversity through identity texts. *Language Teaching Research, 22*(5), 552–569.

Liu, F., & Stapleton, P. (2018). Connecting writing assessment with critical thinking: An exploratory study of alternative rhetorical functions and objects of enquiry in writing prompts. *Assessing Writing, 38*(September), 10–20.

Low, E. L., Hui, C., & Cai, L. (2017). Developing student teachers' critical thinking and professional values: a case study of a teacher educator in Singapore. *Asia Pacific Journal of Education, 37*(4), 535–551.

Manshaee, G., Dastnaee, T. M., Seidi, A., & Davoodi, A. (2014). Comparison of critical thinking in students interested and uninterested in learning a second language. *Theory and Practice in Language Studies, 4*(4), 792–799.

McConachy, T. (2018). Critically engaging with cultural representations in foreign language textbooks. *Intercultural Education, 29*(1), 77–88.

McConachy, T., & Hata, K. (2013). Addressing textbook representations of pragmatics and culture. *ELT Journal, 67*(3), 294–301.

Munandar, M. I., & Ulwiyah, I. (2012). Intercultural approaches to the cultural content of Indonesia' s High School ELT textbooks. *Cross-Cultural Communication, 8*(5), 67–73.

Raigón-Rodríguez, A. (2018). Analysing cultural aspects in EFL textbooks: A skill-based analysis. *Journal of English Studies, 16*,

Shin, J., Eslami, Z. R., & Chen, W. C. (2011). Presentation of local and international culture in current international English-language teaching textbooks. *Language, Culture and Curriculum, 24*(3), 253–268.

Song, B. (2019). Exploring the cultural content in Chinese ELT textbooks from intercultural perspectives. *Journal of Asia TEFL, 16*(1), 267–278.

Thompson, C. C., & Patterson, J. (2019). Promoting critical thinking in an online certificate program for faculty in the health professions. *American Journal of Distance Education, 33*(1), 71–84.

Uzum, B., Yazan, B., & Selvi, A. F. (2018). Inclusive and exclusive uses of we in four American textbooks for multicultural teacher education. *Language Teaching Research, 22*(5), 625–647.

Wiles, J. L., Allen, R. E. S., & Butler, R. (2016). Owning my thoughts was difficult: Encouraging students to read and write critically in a tertiary qualitative research methods course. *Journal of University Teaching and Learning Practice, 13*(1).

Yuan, R., & Stapleton, P. (2020). Student teachers' perceptions of critical thinking and its teaching. *ELT Journal, 74*(1), 40–48.

Post Pandemic L2 Pedagogy – Adi Putra & Arifah Drajati (Eds)
© 2021 Taylor & Francis Group, London, ISBN 978-1-032-05807-8

Building executive function with technological support: Brain-based teaching strategies

Rukminingsih & Januarius Mujiyanto
Universitas Negeri Semarang

Joko Nurkamto
Universitas Sebelas Maret

Rudi Hartono
Universitas Negeri Semarang

ABSTRACT: This study investigated the difference effect of implementing brain-based teaching (BBT) by building executive function (EF) with technology support. A quantitative research design with factorial 2x2 was applied, then two classes involving 38 students of a private college in Indonesia in the fourth grade were chosen as the sample. The instruments used were a questionnaire and a test. Two-way ANOVA was applied to analyze the data. The findings showed that (1) there was distinct impact between experimental class and control class in students' reading achievement, (2) the achievements of the students in the Reading course with high motivation were higher than those with low motivation and (3) there was an interaction between teaching strategies and the level of students' motivation in EFL reading achievement. Educators of ELT have to integrate BBT and EF for the meaningful learning further research and development on different brain-based teaching strategies is suggested.

Keywords: brain-based teaching, executive function, students' motivation level, reading comprehension.

1 INTRODUCTION

Educators have many problems during the COVID-19 Pandemic. They must hold their class full online. They cannot meet their students face-to-face. The teaching and learning processes sometimes have many distractions which are caused by some factors. Students often feel frustrated with fully online learning. Many educators just give many tasks to students without paying attention to the way students learn. As educators, they have to know how learning occurs in students' brains to maximize the learning process (Caine & Caine 1990, 2012; Jensen & Sausa 2001). To reach the optimum teaching process, brain-based teaching strategies by building executive function with technological support was implemented in this study. The executive function is used to actively control the emotions, feelings and actions of students. The executive function influences learning by allowing the students how to organize, manage time and plan in their learning activities. Brain-based teaching strategies by building executive function in this study was implemented by creating a positive environment in the online classroom.

Some studies conducted found that a brain-based learning approach helps teachers to find the way they should teach their students. The emotional basis is the core of the learning process (Rukminingsih 2018; Parr 2016; and Salem 2017) (García et al. 2014). It was also found that certain learning processes, including inferencing and combining prior knowledge with text information during reading, are required to understand the text. Executive function is the management system of the brain. It is associated with students' academic achievements. Some studies dealing with building

 DOI 10.1201/9781003199267-18

executive function found that executive function and the neural network in the brain system can improve the student achievement and motivation (Jacob & Parkinson 2015; Zewelanji et al. 2016; Chevalier et al. 2015).

Therefore, the purpose of this study is to investigate if brain–based teaching strategies by building executive function with technological support can be considered a useful and meaningful learning environment in online learning that could support their reading achievement. This research enables educators and researchers to understand how the student's brain works and learns by building their executive function to contribute to English language learning, especially in EFL reading courses. In addition, we can have a better picture of how foreign language learners can improve and motivate themselves during full online learning, especially in terms of reading courses. Researchers and educators will be able to understand and improve the executive function of students with this knowledge, relevant and innovative pedagogical ideas or teaching methods that allow effective use of brain-based teaching strategies for the achievement of EFL reading. The research questions of this study are:

1. Is there a distinct impact between students who are taught by brain-based teaching strategies by building executive function with technology support with students who are taught by conventional class in students' EFL reading achievement?
2. Do the achievements of students with high motivation in the EFL reading course get higher than those with low motivation?
3. Is there an interaction between teaching strategies and the level of students' motivation in students' EFL reading achievement?

2 LITERATURE REVIEW

2.1 *Brain-based teaching strategies*

Based on some studies above, educators should choose appropriate teaching strategies to cultivate students' motivation. Brain-based teaching strategies by building executive function with technology support can stimulate the students' brain work to learn EFL reading comprehension. Educators in the pandemic COVID-19 era face many problems in teaching by distance education. Brain based teaching is understanding the principles of brain-based learning which involve three instructional techniques which can be implemented in the classroom (Caine & Caine 1994). Three instructional techniques associated with brain-based learning involving orchestral immersion which builds a learning environment which fully engaged students in the class; relaxed alertness which removes fear in the learners while they are learning; and active processing which lets the learners combine and assume materials by actively practicing them (Caine and Caine 1990; Caine et al. 2016).

Students' brains work properly in multifaceted experiences. They need to have various tasks and also teaching strategies. According to Caine and Caine (1994), the brain is unique, learning is changing because it is changing in the brain. To change in long term memory, the brain needs experiences that support the changes which occurred. Those kinds of multifaceted experiences include multisensory input, rewards and motivation, prior knowledge, some examples from concrete to abstract, more practice, telling stories and using computers and other forms of technology. The brain works effectively by seeking the patterns that humans store in their brain by mapping and chunking the information. The information is stored in our brains as patterns. Chevalier (2015) stated keeping the information preserved in the brain is the only way to identify patterns. Teachers need to take fresh data, help students "see" the patterns, associate those patterns with older brain patterns, and generate new ones.

Brain–based teaching strategies generate some meaningful learnings which support the brain work well. Every student has various meaningful learning. What is meaningful to students can be very different from what is meaningful to teachers. Relational memory happens when students are able to connect new learning to something that has occurred in their lives previously. The inclusion of previously stored and mastered patterns or charts makes learning even easier (Willis 2006).

Teachers should provide an environment which lessens stress. A positive classroom environment leads the students to feel safe and comfortable. Lowering stress increases learning. Stressed brains don't learn in the same way as brains that aren't stressed. Students who feel they excel in an area at school will feel better about themselves, and their brains will release chemicals that make them feel good, like dopamine and serotonin, rather than the stress chemical cortisol. Cortisol is elevated by stress, but the amount of cortisol is not specifically related to the effects of memory stress. This means that if students get stress in their learning process, their cortisol will be increased (Shields 2017).

2.2 *Executive function with technology support*

Executive function is a top-down monitoring and control process which activates the learners' behavior (Diamond & Ling 2016). Inhibition (control of one's actions, attention, thoughts and emotions), working memory (temporarily retaining and using information) and cognitive flexibility (effectively switching between tasks) are the key executive functions (Diamond & Ling 2016; Miyake et al. 2000; Zelazo et al. 2013). Another brain-based teaching strategy is building students' executive function. The executive function is an extraordinary capacity to control the emotions, feelings and behaviors of humans actively in order to accomplish goals.

Neuroscience helps the educators' understand the students' brain work, their strengths and weaknesses so it can help teachers understand them. The brain is plastic and can be formed, altered and activated. Research has shown that when students realize and believe that they can change their brains and grow their intellect, they work harder, persist through difficulty and attain greater achievement. (Rukminingsih 2018 and Parr 2016). Brain based teaching strategies to help build executive function developed by Caine & Caine (1990, 2012), Jensen (1996), Sousa (2001) and Chevalier (2015). Certain areas of the brain can be developed by activating students' executive function which are responsible for working memory and emotional control.

As the brain grows and develops, it is ready for various kinds of learning. Their brains have adapted quite well to the high-tech world, and pandemic COVID-19 forces the education to implement full online learning. The use of technology is implemented in all schools in Indonesia. Adult learners utilize inhibition to rally their attention with various tasks in various online platforms both synchronous and asynchronous such as using Google classroom, telegram and zooming. Working memory involves the collection and retrieval of information at the same time. In reading comprehension, for instance, students have to trigger the content schemes of students on the same topic with the text they read, bring context knowledge to the forefront of their memory, use all the details to easily understand the text. Cognitive versatility allows students to support active learning. Cognitive flexibility enables students to make agreement with the teacher or lecturer in their classroom activities while course outline is made (Zewelanji 2016, and Chevalier 2015)

3 RESEARCH METHOD

This study was quantitative factorial because it had two factors and each factor had two levels, participants 2 × 2 factorial design (Ary et al. 2010). The sample was taken from students who were taking Critical Reading course from two different classes of an English department in one of private college in Indonesia. The participants were 76 students involving 38 students for the experimental group and 38 students for the control group. There are three variables in this study, namely two independent variables (brain-based teaching strategies by building executive function with technology support and online instruction with flipped classroom as a conventional teaching strategy. Then the moderator variable was students' reading motivation and the dependent variable was students' reading achievement.

The instruments used in this study were motivation reading questionnaire and reading comprehension test. The questionnaire was used to measure students' reading motivation level to classify students into high and low levels of reading motivation. The questionnaire with Likert scale in which the questionnaire was designed with related indicators of students' reading motivation. The reading motivation questionnaire aimed to classify students with high and low levels of reading motivation. Reading comprehension test was used to assess students' achievement in EFL reading comprehension. Two-way Analysis of Variance (ANOVA) at the level of significance alpha = 0.05 was the data analysis used in this study. It was used to test the three hypotheses. There were two assumption requirement of the two-way ANOVA, namely the normality and homogeneity of the test should be met. Normality was evaluated using the Lilliefors test and homogeneity was tested using the F test and Barlet test.

4 FINDINGS AND DISCUSSION

The findings are presented in two sections to answer the research questions. First, the summary of data description is presented in Table 1 and the second, Summary on calculation result of two -way ANOVA data is presented in Table 2.

The following is the summary of the two-way ANOVA computation which contained the variance related to the score of means, teaching strategies, students' motivation, interaction, error, and means

Table 1. Summary of data description.

Statistical Values	A1	A2	B1	B2	A1B1	A1B2	A2B1	A2B2
N	38	38	38	38	19	19	19	19
Highest score	37	34	93	71	37	29	33	29
Lowest score	20	20	20	20	20	20	20	20
Mean	28.08	26.89	81.63	63.68	31.63	24.58	26.58	25.58
Median	28.00	27.50	80.00	64.50	31.00	25.00	26.00	26.00
Mode	28.00	28.00	76.00	64.00	30.00	27.00	32.00	29.00
Standard deviation	4.54	4.09	5.62	4.39	2.73	3.06	4.35	3.06
Variance	20.57	16.69	31.59	19.25	7.47	9.37	18.92	7.37

Notes:
A1: group of students taught by using brain-based teaching strategies by building executive function with technology support
A2: group of students taught by using online instruction integrated with flipped classroom
B1: group of students with high motivation
B2: group of students with low motivation
A1B1: group of high motivation taught by using brain -based teaching strategies by building executive function with technology support
A1B2: group of low motivation taught by using brain -based teaching strategies by building executive function with technology support
A2B1: group of high motivation taught by using online instruction integrated with flipped classroom
A2B2: group of low motivation taught by using online instruction integrated with flipped classroom

Table 2. Summary on calculation result of two-way ANOVA.

Variance	**Dk (Df)**	**Sum of squares**	**Mean square**	**F observed**	Ft A(α) = 0.05
Teaching strategies	1	308	308	28.12	3.88
Students' motivation	1	78	78	7.82	3.88
Interaction	1	174	174	15.22	3.88
Error	72	812	11.27	–	–
Means of treatment	1	557155	–	–	–
Total	76	57155	–	–	–

of treatment. By looking at this description of the analysis of variance, it is easier to take into account the analysis related to two-way ANOVA as shown in the following.

This description of the two-way ANOVA measurement results was used to validate or identify the research hypotheses. The table above described the result of the testing hypothesis. Based on the data on the table above, it was concluded that the alternative hypotheses were confirmed.

The value of observed F exceeds the value of F from table in the three variances (teaching strategy (28.12), motivation (7.82), and interaction (15.22) whereas the value of F from table was merely 3.88 for three variances. It could be seen that the three hypotheses were confirmed at alpha 0.05, as the first hypothesis is that the students' achievement in reading comprehension taught by using brain-based teaching strategies by building executive function with technology support was higher than those taught by using online instruction integrated with flipped classroom strategy was confirmed; the second hypothesis is that the achievements of the students in the reading course with high motivation were higher than those with low motivation was confirmed; the third hypothesis is that there was an interaction between teaching strategies and the level of students' motivation in EFL reading achievement was confirmed.

Based on the finding, it can be concluded that students' achievement in Reading Comprehension by employing brain-based teaching strategies by building executive function, there is a great increase in conceptual understanding after the implementation of online based- instruction based on brain-based teaching in Reading comprehension class. Pourhossein (2014) and Chevalier (2015) stated that using technology can create a learning atmosphere centered around the learner rather than the teacher that in turn creates positive changes. Caineet al. 2016; Rukminingsih 2018; and Zewelanji 2016) have stated that building executive function can lead the students to change their brains and grow their intelligence. They work harder, increase their spirit and make greater achievement in their reading comprehension.

The strategies of brain-based teaching by building executive function with supporting technology involve: (1) creating a positive emotional environment in online classroom, (2) providing opportunities to apply learning, (3) introducing activities to support developing executive function and prior knowledge activation and transfer opportunities and (4) employing model higher thinking skill which adapted from Caine et al. 2016; Sousa 2001 and Handayani et al. 2020.

By employing online instruction, such as in Google classrooms, teachers can use feedback loops to find out whether the students' perception matches their expectation. This step is used to organize information in the brain at different levels. Students must transform information as their own learning with the use of working memory and prior knowledge to form long-term allow students to use the information into different products that can become a trigger for conceptual understanding. Handayani et al. (2020); Ramakrishnan andAnnakodi (2013) and Rukminingsih (2018) found that teachers should make use of brain-based teaching strategy and the concept of brain-based learning in the classroom.

5 CONCLUSION

Based on findings and discussion that brain-based teaching by building executive function had a statistically significant influence on the students' reading achievement and motivation. Based on finding and discussing, brain-based teaching strategies by building executive function with technology support had three conclusions. The students' achievement in reading comprehension taught by using brain-based teaching strategies by building executive function with technology support is effective. There is a statistically significant influence on the students' motivation and achievement in Reading comprehension. There is an interaction between teaching strategy and students' motivation level in reading comprehension.

The findings of this study confirms existing empirical evidence dealing with brain-based teaching strategies which have been implemented by building students' executive function with technological support in one of private college EFL students in Indonesia. These findings have the following pedagogical implications for current and prospective English teachers and lecturers, students and educational authorities it is great significance that to provide teachers and lecturers with some knowledge in which they can make the students motivated and get a good achievement in full online learning by considering how their brains work and learn by implementing brain-based teaching strategies. Building students executive function stimulates students to have good emotion, positive climate in an online class environment and strengthen their cognitive skills especially in EFL reading courses. In foreign/second language learning, there is an increasing need to understand more about brain-based teaching strategies; therefore, it is hoped that this research will contribute to the further knowledge about students in EFL reading classrooms and facilitate more studies on brain-based teaching strategies with technological support.

REFERENCES

Ary, D., Jacobs, L. C., & Sorensen, C. (2010). *Introduction to Research in Education.* (8th ed.). Wadsworth.

Chevalier, N., Martis, S. B., Curran, T., & Munakata, Y. (2015). Metacognitive processes in executive control development: The case of reactive and proactive control. *Journal of Cognitive Neuroscience, 27*(6), 1125–1136.

Caine, R. N., & Caine, G. (1990). *Understanding a brain-based approach to learning and* teaching. (1st ed.). Press, Inc.

Caine, R.N., Caine, Geoferry, McClintic, Carol & Klimek, K.J. (2016). *Brain mind learning* principles *in action.* (3rd ed.). Corwin.

Diamond, A., & Ling, D. S. (2016). Conclusions about interventions, programs, and approaches for improving executive functions that appear justified and those that, despite much hype, do not. *Developmental Cognitive Neuroscience Journal, 18*(2016), 34–48.

García-Madruga, J. A., Vila, J. O., Gómez-Veiga, I., Duque, G., & Elosúa, M. R. (2014). Executive processes, reading comprehension and academic achievement in 3th grade primary students. *Learning and Individual Differences Journal*, 35, 41–48.

Handayani, S.B., Aloysius Duran Corebima, D.A, Susilo, H. and Mahanal,S. (2020). Developing brain-based learning (bbl) model integrated with whole brain teaching (wbt) model on science learning in junior high school in Malang. *Universal Journal of Educational Research* 8(4A): 59–69, 2020.

Pourhossein Gilakjani, A. (2014). A detailed analysis over some important issues towards using computer technology into the EFL classrooms. *Universal Journal of Educational Research, 2*(2), 146–153.

Rukminingsih. (2018, October 13–15). *Integrating neurodidactics Stimulation into Blended Learning in accommodating Students English Learning in EFL Setting* [Paper presentation].13th Annual Asian Conference Education, Tokyo, IAFOR, Japan.

Ramakrishnan, J. and Annakodi. (2013). Brain based learning strategies. *International Journal of Innovative Research and Studies. 2*(5).235–242.

Salem, S.M.A.F. (2017). Engaging ESP students with brain-based learning for improved listening skills, vocabulary retention and motivation. *English Language Teaching Journal, 10*(12), 144–154.

Sousa, D. A. (2001). How the brain learns: A classroom teacher's guide. California, Corwin Press, Inc.

Parr. Tara, L. (2016). A Brain-targeted teaching framework: Modeling the intended change in professional development to increase knowledge of learning sciences research and Influence pedagogical change in K-12 public classrooms. Dissertation. Doctor of Education Field of Educational Leadership and Management. Drexel University.

Zewelanji, N. Serpell and Alena, G. and Esposito. (2016). Development of executive functions: implications for educational policy and practice. *Policy Insights from the Behavioral and Brain Sciences Journal*, 3(2), 203–210.

Post Pandemic L2 Pedagogy – Adi Putra & Arifah Drajati (Eds)

Engaging EFL learners of English writing with prewriting activities assisted technology

Dwi Sloria Suharti & Eka Ugi Sutikno
Universitas Muhammadiyah Tangerang, Indonesia

Hani Dewi Aries Santi
Universitas Esa Unggul, Jakarta, Indonesia

ABSTRACT: Using Assisted Technology in Prewriting activities to teach English writing to EFL students at the university level is not a new venture in TEFL. However, this learning stage still rarely applies in Indonesia. To fill this gap, we implemented Prewriting Activities Assisted Technology to engage EFL learners in the prewriting activities using assisted technology. This article purposely describes our experience using Prewriting Activities Assisted Technology as a teaching innovation with EFL learners of 20-23. In this project, students employed Prewriting Activities Assisted Technology as their endeavors to compose EFL writing. In this matter, they utilized different prewriting activities with technological assistance. The prewriting activities were readings, watching movies, discussions, brainstorming, webbing, and outlining. The Prewriting Activities Assisted Technology project is a practical suggestion for EFL writing teachers to engage EFL learners in significant project-based language learning.

Keywords: EFL learners, English writing; prewriting activities assisted technology; project-based language learning

1 INTRODUCTION

Derakhshan and Karimian Shirejini (2020) identified some Iranian EFL students' perceptions of the most common sources of writing difficulties. The interviews indicated that grammar, spelling, punctuation, paragraph coherence, organization, words, and rhetorical structures make the writing task difficult. Moreover, the lack of grammar or vocabulary makes it difficult for EFL language learners to generate a good paragraph composition (Hossain 2015). The attributes of writing that provide it such significance and various factors that were causing writing one the most challenging language skills to learn have been counted overhead (Javadi-Safa 2018).

However, the importance of writing ability is undeniable in second language teaching and research (Javadi-Safa 2018). Moreover, EFL learners' achievement in English writing passes their profits not only in their English learning but also in their life professions (Tuan 2010). Nevertheless, language learners viewed it as a troublesome aptitude, particularly in EFL contexts, where students encounter many challenges in composing (Fareed et al. 2016). Mother-tongue intervention is closely related to English writing skills (Ibrahim & Ibrahim 2018). Weigle (2014) asserted that EFL writers could not transmit their logical information from their first language to English. Lack of motivation also makes EFL students unable to compose English compositions (Long et al. 2013). The inadequacy of reading even loses the idea of writing English (Hidayati 2018). EFL writing teachers must help students produce text using English appropriate for vocabulary and grammar. To write, the EFL learner needs a process to develop this skill. EFL writers need a strategy of writing for the process of writing English.

 DOI 10.1201/9781003199267-19

Writing English in the writing process is a dynamic task (Toraskar & Lee 2016). Particularly in prewriting, EFL learners often find themselves without writing ideas (Abderraouf 2016; Muslim 2014). Strategies or techniques of EFL teaching-learning are necessary. Learning techniques encourage EFL learners to learn English in initial writing tasks, such as prewriting. Prewriting activities can make it easier to write English (Mogahed 2013). Xu et al. (2018) suggest that technology innovations could improve English language learners' (ELLs)' writing efficiency. Their study discusses crucial elements regarding how adult ESL/EFL can use technologies to increase their writing content, and the impact of instructional interventions may be made more significant.

Research on using technology-assisted in EFL teaching Indonesia environment prewriting is limited. Therefore, this research investigated how students apply a prewriting technique using technology-assisted learning EFL writing (Prewriting Activities Assisted Technology project). Discovering the technology-assisted tools/applications used. Besides, recognizing the advantages and constraints found.

2 LITERATURE REVIEW

EFL students find classroom activities are demotivating and uninteresting (Mahboob & Elyas 2014). Outmoded teaching techniques are often routine, which may be a reason for students' decreased performance levels and lack of interest. Thus, innovative approaches are required to address students' expectations while stimulating their engagement and enhancing learning opportunities. Research into Technology Assisted Language Learning (TALL) needs is crucial if we hope to have better teaching insights. Ko (2017) believed that TALL has emerged as a powerful way of enhancing language learning classes. The latest electronic gadgets, which have virtually impacted all human life features, are Internet-skilled mobile telephones (Ahmad 2016). Microsoft Word has beneficial effects on EFL teachers' and learners' grammar and enhancement skills (Castrillo de Larreta-Azelain & Martín Monje 2016; Salehi & Amiri 2019).

In previous studies, Al-Bogami and Elyas (2020) asserted that student engagement is a dynamic term that can be viewed differently. Researchers recognized engagement as a multidimensional construct. Students' engagement is related to classroom instruction. Affective engagement refers to students' emotional involvement, while behavioral engagement refers to students' engagement in learning. Then, to maximize students' engagement in innovative, interactive, critical, and communicative learning activities, EFL teachers could deploy mobile technologies (Traxler 2010). The study reported that students learn better if they are fully engaged and interested in the class. They could produce, consume, and store content and conversation by utilizing their handheld devices. Employing mobile learning (m-Learning) with tablet gadgets and apps could help develop independent use and autonomous engagement.

Moreover, technology and personal interests are used to encourage language learning outside the classroom (Nguyen & Terry 2017). The online learning environment helps learners track and engage in their learning process (Lim 2004). EFL teachers can also acquire new teaching strategies with various technological tools (Zandi & Krish 2017). Angelaina and Jimoyiannis (2012) reported that blogs are a modern platform for students to promote contact, engagement, teamwork, and mutual knowledge. The study reveals that blogs help students learning from text to deeper comprehension and knowledge construction in the EFL writing context. The teachers applied Blogs in educational settings for their interactive teaching attributes. The blog contribution and sharing in class can help students grow collaboratively. The students used them to keep personal journals. The study indicated the potential of blog-based learning practices as a formative assessment method.

Lan et al. (2015) stated that computer-supported cooperative learning (CSCL), prewriting techniques influence new EFL students' English writing efficiency and motivation. EFL teachers considered the various impacts of using different beginning writers' strategies from other aspects to enhance learning effects. The online learning environment helps learners track and engage in their learning process. Moreover, McDonough and De Vleeschauwer (2019) reported that smart learning prewriting might be scaffolded by offering specific guidance and visual tools. Consistency benefits are obtained from collective experience. Weigle (2014) believed those prewriting activities could target linguistics development, fluency, and idea generation. They help EFL learners build up knowledge about a topic. Activities such as freewriting, generating lists of ideas associated with a topic, and making a mind map of related concepts are appropriate at all proficiency and experience levels. Additional activities may include targeted lessons on specific writing aspects, including using dictionaries or other reference materials in writing. Furthermore, prewriting activities also promote the identification of rhetorical structures of texts, such as structured language practice, readings, watching films, discussions, brainstorming, webbing, and outlining.

EFL teaching-learning ran online setting, EFL teaching-learning activities assisted technology became a paramount need. Using Assisted Technology in Prewriting activities to teach English writing to EFL students at the university level is not a new venture in TEFL. However, this learning stage is still rarely applied in Indonesia. To fill this gap, we implemented Prewriting Activities Assisted Technology to engage EFL learners. In teaching English writing to EFL students, the teacher could deploy Prewriting Activities Assisted Technology. This article describes our experience using Prewriting Activities Assisted Technology as an innovation for students at a private university in Indonesia.

3 RESEARCH METHOD

The researchers examined the students' engagement employing the prewriting activities theory proposed by Weigle (2014). This teaching strategy (prewriting activities: language practice, readings, movies, discussions, brainstorming, webbing, and outlining) are analyzed to know the assisted technology/tools applied by students in their prewriting activities. The project's practical model is for EFL writing teachers to experience Prewriting Activities Assisted Technology. Thus, one of the researchers who was a participant-observer applied this project to her students. It is to engage the learners in this significant project-based language learning. Seventy-two participants were recruited, 61 female and 11 male, from 110 students (aged 20-23 years) who enrolled in a Journalism course for undergraduates semester six at a private university in Indonesia. Then, observation and questionnaire were deployed to gain the data. The participants used the assignments on the project to analyze their engagement in project-based language learning. The participants employed Prewriting Activities Assisted Technology to write EFL English. First, they were asked to do the tasks in the Microsoft office word file. Then they posted the works on their blogs.

The lessons started before the pandemic, with six meetings using blended learning, namely synchronous meetings in the classroom and assignments in google classroom and WhatsApp Group (WAG). It took four weeks for an introduction to the lessons and two weeks for practice. In this project, the participants undergo a full online Journalism lesson: News Story, Editorial, Advertisement, Review a Movie, and Broadcasting News. The EFL Journalism writing tasks with the project are in Table 1.

4 FINDINGS AND DISCUSSION

The researchers found that The Prewriting Activities Assisted Technology project proposed by Weigle (2014) provided students with a new writing experience in writing their EFL Journalism writing favorably. Students engaged in classroom activities with an average of tools in their prewriting activities were webbing, reading, and summarizing the news text. The assisted technology used was a mobile phone, notebook (laptop), Google Translate, Microsoft Word, Grammarly, YouTube, Videomaker app, and Anchor/ Spotify app. The study also reported that Blogs were selected tools besides could help students learn from text to deeper understanding and knowledge construction in the EFL writing context and one way of assessment technique and promote contact, engagement, teamwork, and mutual knowledge. In line with Angelaina and Jimoyiannis (2012), it is believed that blogs are a modern platform for students to promote contact, engagement, teamwork, and mutual knowledge. The study reveals that blogs help students learning from text to deeper comprehension

Table 1. Project-based EFL journalism writing tasks.

Project-based EFL Journalism writing tasks	Directions	Prewriting activities and tools used.
Task 1. Editorial Cartoon	Select a news text, write the source and the link you retrieved, and analyze the editorials' type. Write also references	Brainstorm: webbing (search engine), reading e-books and online resources, a mobile phone, notebook (laptop), Google Translate, Microsoft Word, Grammarly
Task 2. Review a Movie	Analyze a recent movie that you are interested in reviewing by mind mapping. After you analyzed the movie, write a review of the movie using the information above.	Watch movies: YouTube, movie streaming websites, a mobile phone, notebook (laptop), Google Translate, Microsoft Word, Grammarly.
Task 3. News Story	After you select your photos you are interested in, create a writing caption. Write a news story which is including the elements of a news article.	Observation/ interview using a mobile phone, notebook (laptop), Google Translate, Microsoft Word, Grammarly
Task 4. Advertisement News	Before writing advertisement news, structure a survey of a product that you want to advertise. Then survey your customers/readers. Identifying the excellence of your product gets some testimonials.	Do a little survey using a mobile phone, notebook (laptop), Google Translate, Microsoft Word, Grammarly.
Task 5. Broadcasting News	Students must make an overall plan and layout for a print publication. There must also be a precise plan for a broadcast program, including a news broadcast.	Webbing, reading, and summarizing the news text using a mobile phone, notebook (laptop), Google Translate, Microsoft Word, Grammarly, YouTube, Videomaker app, and Anchor/ Spotify app

(Source: Prepared by the authors, 2021)

and knowledge construction in the EFL writing context. In short, these are some of their works in their blogs:

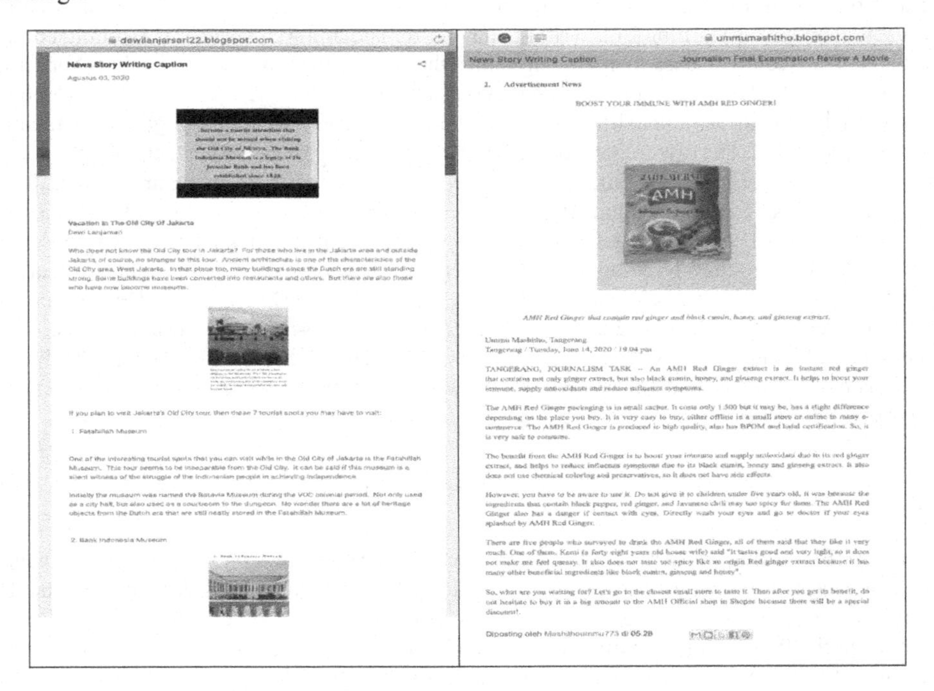

Figure 1. Students' blog examples.

Moreover, students participated as a critical thinker, as Traxler (2010) believed. They made decisions about the person who would like them interviewed, selected the news text, selected the observed place. They could voice their ideas based on the photographs they took. In short, it builds both affective and behavioral engagement, as affirmed by Al-Bogami and Elyas (2020). Besides it maximizes students' engagement in critical thinking skills, it is also assisted student-centered learning, learners' autonomy, and motivation to develop successful foreign language skills, as Traxler (2010) asserted. The students' engagement is depicted in their EFL writing task on their blogs. It is in line with the study done by Angelaina and Jimoyiannis (2012). They asserted that blogs are a new way for learners to engage and learn from one another.

Students collected the source of writing as their prewriting activities. Observation as a prewriting activity, students used smartphones and notebooks. They employed video calls and social media to ask interviewees. Furthermore, prewriting activity: Brainstorming, students applied to the web. The most used tools used were Google and Mozilla Firefox. They also did watch movies. They used DVDs or CDs to watch movies. Most of them used YouTube. Others used movie streaming websites like Netflix, Viu, iflix, Maxstream, hooq, and indoxx1. They watched a movie, or short video lessons affected students' language skills, particularly EFL writing skills, as Castrillo de Larreta-Azelain and Martín Monje (2016) also believed. They felt practical and became autonomous learners to get information about the topics they wanted to write. As Traxler (2010), it is a line that assisted technology enables students to be autonomous learners. It enhances language learning experiences.

In the prewriting activities, they also did the reading. Using e-books and online resources was significant because they did not use many papers. EFL students engage in their EFL language learning through the prewriting activities assisted by technology. It is similar to Lim (2004), Nguyen and Terry (2017), McDonough and De Vleeschauwer (2019), and Zandi and Krish 2017) that a reformation in the teaching using various tools as assisted technology could support the students' engagement in their learning activities.

Moreover, by employing assisted technology in the EFL prewriting activities, the students had the best opportunity to generate their ideas, as asserted by Mogahed (2013). In line with Ko (2017), this Prewriting Activities Assisted Technology as TALL is a powerful technique of developing language learning classes. The researchers confirmed that prewriting activities could make it easier for EFL learners to write in English. It helps them write their main paragraph idea anytime, anywhere, so it is useful and makes learning EFL writing in a non-threatening way because they could edit their draft text. They could discuss their writing topics through applications such as WhatsApp. It is in line with Ahmad (2016). Ahmad (2016) believed that gadgets, Internet-skilled mobile telephones, could virtually impact all human life features, including students' learning engagement in their EFL prewriting activities. The researchers found that the students experienced that technology is helpful, adaptable, and utilized whenever and anyplace. It could also correct their grammar structure. It helps them start writing EFL quickly and simplifies their writing, as Salehi and Amiri (2019) asserted. He said that a word processor could function to fix grammatical and spelling errors and quicker writing. Finally, it is smart learning. Prewriting activities proposed specific guidance, visual tools, and benefits from collective experience, which is in line with McDonough and De Vleeschauwer (2019). However, some students did not encounter any difficulties in employing assisted technology to compose EFL during prewriting. Consequently, the Prewriting Activities Assisted Technology project is motivating as students are actively engaged in prewriting practices, which are essential for writing their EFL.

5 CONCLUSION

Engaging EFL writing learners with Prewriting Activities Assisted Technology is a time expending endeavor, but by deploying this project, the students became engaged as critical thinkers. They felt realistic and became autonomous learners to study the subjects they needed to publish. For this reason, the Prewriting Activities Assisted Technology project is optional. The implications of this study, the project designed, makes students actively engaged in prewriting practices, which are essential for their EFL writing process. It maximizes students' engagement in critical thinking skills, student-centered learning, learners' autonomy, and motivation to develop their prewriting activities, such as discussions, brainstorming, observing, outlining, interviewing, surveying, webbing, reading, and summarizing the news text. They did not encounter any difficulties in employing assisted technology such as a mobile phone, notebook (laptop), Google Translate, Microsoft Word, Grammarly, YouTube, Videomaker app, and Anchor/Spotify app to compose EFL. On another pedagogical reminder, EFL writing teachers should inspire non-autonomous learning students. Moreover, the students may obtain adequate training on the project. The study of the overall writing process activities applying assisted technology is recommended for further studies.

ACKNOWLEDGEMENT

Our gratitude is to the Kemenristek/BRIN 2020 for financial support. Furthermore, we are also indebted to the Research Manager of Universitas Muhammadiyah Tangerang (LP3M) for supporting us. So, our research ran well.

REFERENCES

Abderraouf, A. (2016). Writing difficulties and common errors in writing: a case study of third-year LMD students of English at The University of Bejaia. In the *University of Bejaia*. The University of Bejaia.

Ahmad, J. (2016). Technology-assisted language learning is a silver bullet for enhancing language competence and performance: A case study. *International Journal of Applied Linguistics and English Literature, 5*(7 Special Issue), 118–131.

Al-Bogami, B., & Elyas, T. (2020). Promoting Middle School Students' Engagement Through Incorporating iPad Apps in EFL/ESL Classes. *SAGE Open, 10*(2).

Angelaina, S., & Jimoyiannis, A. (2012). Analyzing students' engagement and learning presence in an educational blog community. *Educational Media International, 49*(3), 183–200.

Castrillo de Larreta-Azelain, M. D., & Martín Monje, E. (2016). Students' engagement in online language learning through short video lessons. *Porta Linguarum, 2016*(26), 177–186.

Celce-Murcia, Marianna. (2014). An Overview of Language Teaching Methods and Approaches. In Marianne Celce-Murcia, D. M. Brinton, & M. A. Snow (Eds.), *Teaching English as a second or foreign language* (4th Edition, pp. 2–14). National Geographic Learning, HEINLE Cengage Learning.

Derakhshan, A., & Karimian Shirejini, R. (2020). An Investigation of the Iranian EFL Learners' Perceptions Towards the Most Common Writing Problems. *SAGE Open, 10*(2), 1–10.

Fareed, M., Ashraf, A., & Bilal, M. (2016). ESL Learners' Writing Skills: Problems, Factors and Suggestions. *Journal of Education & Social Sciences, 4*(2), 83–94.

Hidayati, K. H. (2018). Teaching Writing to EFL Learners: An Investigation of Challenges Confronted by Indonesian Teachers. *Langkawi: Journal of The Association for Arabic and English, 4*(1), 21.

Hossain, M. I. (2015). *Teaching Productive Skills to the Students: A Secondary Level Scenario*. BRAC University, Dhaka, Bangladesh.

Ibrahim, A., & Ibrahim, S. (2018). The first language influence on the EFL-learners ' writing performance?: Errors analysis and remedial perspective. *Journal of Education and Practice, 9*(14), 152–165.

Javadi-Safa, A. (2018). A Brief Overview of Key Issues in Second Language Writing Teaching and Research. *International Journal of Education and Literacy Studies, 6*(2), 15–25.

Ko, M. H. (2017). Learner perspectives regarding device type in technology-assisted language learning. *Computer Assisted Language Learning, 30*(8), 844–863.

Lan, Y. J., Sung, Y. T., Cheng, C. C., & Chang, K. E. (2015). Computer-supported cooperative prewriting for enhancing young EFL learners' writing performance. *Language, Learning and Technology, 19*(2), 134–155.

Lim, C. P. (2004). Engaging Learners in Online Learning Environments. *TechTrends, 48*(4), 16–23.

Long, C., Ming, Z., & Chen, L. (2013). The study of student motivation on English learning in Junior middle school – A case study of No.5 middle school in Gejiu. *English Language Teaching, 6*(9), 136–145.

Mahboob, A., & Elyas, T. (2014). English in the Kingdom of Saudi Arabia. *World Englishes, 33*(1), 128–142.

McDonough, K., & De Vleeschauwer, J. (2019). Comparing the effect of collaborative and individual prewriting on EFL learners' writing development. *Journal of Second Language Writing, 44*(January), 123–130.

Mogahed, M. M. (2013). Planning out prewriting activities. *International Journal of English and Literature, 4*(3), 60–68. https://doi.org/10.5897/IJEL12.120
Muslim, I. M. (2014). Helping EFL Students Improve their Writing. *International Journal of Humanities and Social Science, 4*(2), 105–112.
Nguyen, H., & Terry, D. R. (2017). English Learning Strategies among EFL Learners: A Narrative Approach. *IAFOR Journal of Language Learning, 3*(1), 4–19.
Salehi, H., & Amiri, B. (2019). Impacts of Using Microsoft Word (MS) Software on Iranian EFL Lecturers' Grammar Knowledge. *International Journal of Research in English Education, 4*(1), 1–10.
Toraskar, H. B., & Lee, P. K. L. (2016). Hong Kong Undergraduate Students' Academic Writing: 21st Century Problems, Solutions and Strategies. *The Journal of Asia TEFL, 13*(4), 372–380.
Traxler, J. (2010). Will Student Devices Deliver Innovation, Inclusion, and Transformation? *Journal of the Research Center for Educational Technology (RCET), 6*(1), 3–15.
Tuan, L. T. (2010). Enhancing EFL Learners' Writing Skill via Journal Writing. *English Language Teaching, 3*(3), 81–88.
Xu, Z., Banerjee, M., Ramirez, G., Zhu, G., & Wijekumar, K. (Kay). (2018). The effectiveness of educational technology applications on adult English language learners' writing quality: a meta-analysis. *Computer Assisted Language Learning, 32*(1–2), 1–31.
Zandi, P., & Krish, P. (2017). Enhancing EFL writing instruction through technology in Iran. *Research on Humanities and Social Sciences, 7*(18), 75–83.

Post Pandemic L2 Pedagogy – Adi Putra & Arifah Drajati (Eds)
 ISBN 978-1-032-05807-8

Integrating digital teaching package for the German language classroom

Dwi Putri Ningsih & Sonya Puspasari Suganda
Universitas Indonesia, Depok, Indonesia

ABSTRACT: This paper intends to examine the perceptions of teachers and students of the use of *Digitales Unterrichtspaket* or Digital Teaching Package (DUP) in the German language online learning. In this study, a total of 20 students and three teachers are involved. A questionnaire consisting of closed and open-ended questions was distributed to the students and teachers. Data analysis was conducted by examining the results of descriptive statistics and by examining the participants' open-ended responses. The results showed that students and teachers perceived DUP as an attractive and useful tool for online learning. Teachers do not have to spend a lot of time preparing instructional materials and students are more involved in the learning process. DUP facilitates interactions that lead to successful learning. These findings help teachers evaluate their instruction with DUP during the pandemic.

Keywords: Digital teaching package, German language, online learning, perceptions

1 INTRODUCTION

Online learning refers to learning delivery formats using technological tools and the internet. However, students tend to feel disconnected, isolated, distracted and lack personal attention in online learning. On other hand, the more students interact and engage in the classroom, the more likely they are to persevere in learning. Consequently, teachers must maintain connectivity in the online classroom through interaction. Interaction can be made through the use of appropriate instructional tools. Digital tools are widely used in L2 learning because digital tools offer teachers and students teaching and learning resources. Furthermore, digital tools help teachers to create differentiated L2 learning instruction (Hover & Wise 2020). Later, digital tools can be seen as effective tools in the context of L2 online learning.

In the last ten years, there have also been studies investigating the use of digital tools in online learning that help improve academic achievements. A study by Thongsri (2019) showed that the students' positive perception of digital tools usage later helps students to achieve academic success. Another study by Hromalik and Koszalka (2018) showed that teachers see digital tools as useful tools because they help students to improve their L2 performance in the classroom. There was also a study that highlighted issues regarding the perceptions of digital tools used in online learning (Bueno-Alastuey & Pérez 2013). In general, all these studies reported that positive perceptions of digital tools help students to achieve learning success and teachers' positive perceptions can improve their instructional design in online learning.

During the COVID-19 pandemic, instructional setting shifts from face-to-face to fully online. This can influence the perceptions of students and teachers regarding the use of digital tools. This study aims at finding out whether, during a pandemic, perceptions of the use of digital tools change. The digital tools being used are the digital teaching package or Digitales Unterrichtspakette (DUP)

DOI 10.1201/9781003199267-20

2 LITERATURE REVIEW

Online learning is aligned with the learning delivery and experience that students would not rely on their physical or virtual location and learning using technologies as well as the internet in a synchronous or asynchronous setting (Singh & Thurman 2019). In online learning, teachers need to encourage the learners to be active, because there is a tendency of learners and instructors to feel disconnected. Besides, social presence and interaction play an essential role in creating an interactive relationship with others. Digital tools are seen as useful tools in the context of online L2 learning. Teachers need to choose and use immersive and enjoyable for students because it helps to enhance students' self-regulation, performance, achievement and learning design in online learning (Hromalik & Koszalka 2018).

Digital tools support the learning process by facilitating various practices depending on the learner's goals, abilities, and interests the powerful digital instructional tool in the context of L2 learning reflects four main purposes and benefits: 1) technologies promote the practice, 2) deliver learning content, 3) facilitate interaction and 4) restructure teaching methods (Zhang & Zou 2020). The importance of using appropriate digital tools in online learning affects the outcome of L2 learning. As Sun and Chen (2016) mentioned that digital tools help learners and instructors improve their competencies in the learning process, which leads to successful online learning. However, student's and teachers' perceptions and readiness of digital tools in online learning are also important, because it affects the L2 learning outcomes and also the course satisfaction (Wei & Chou 2020).

In the context of online German language learning, one of the digital tools used is *Digitales Unterrichtspaket* (DUP), or digital teaching package. DUP is an interactive digital tool, which consists of all the contents of the printed textbook available to learners via screen sharing and even plays audio and video directly (Ernst Klett Sprachen GmbH 2020). DUPs interactive tools and contents facilitate the interactions (students-students, student-teachers, and students-learning material) and also promotes listening, speaking, reading, and writing exercises. Teachers' and students' perceptions of DUP usage need to be addressed because a positive perception of instructional tools promotes learner's and teacher's perception through the ease of use, the flexibility of instructional tools, and learning activities, which can help to make effective online language learning. (Thongsri et al. 2019; Wei & Chou 2020). Findings from the studies above revealed that the perception of digital use can help to achieve L2 online learning in an ideal learning setting. Therefore, I would like to examine how teachers and students perceive digital tools during the COVID-19 pandemic.

3 METHOD

The participants were female students from an intensive German class in the 10th-grade (N = 20) and the age range of participants is between 14–15 years. There are only twenty students that completed the consent form and completed the questionnaire. In this research, there are also teachers as participants (N = 3) who teach in an intensive German class in a private senior high school in a metropolitan area in Indonesia. During the research, all participants were having the experience to use DUP for three months in online L2 learning. This study used a mixed-method approach and the survey research design to find out student's and teachers' perceptions toward the use of DUP and the benefit of DUP implementation in online learning. To collect the data, questionnaires were distributed online to students and teachers. The questionnaire's results aim to get a snapshot of the participant's perception of current DUP usage during the pandemic.

Each questionnaire consisted of eleven closed-ended questions, which derived from Zhang & Zhou's (2020) work the four main purposes and advantages of technologies as teaching tools: (1) Promotes practices – Items Q8, Q9, Q10, and Q11, (2) Deliver learning content – Items Q1 and Q2, (3) Facilitate interaction – Items Q3, Q4, and Q5, (4) Restructure teaching methods – Items Q6 and Q7, and three open-ended questions. The questionnaires used four Likert scales (1 – strongly disagree, 2 – strongly disagree, 3 – agree, 4 – strongly agree) and three items with open-ended questions. Results were analyzed using the mean values of the participants' responses. The mean interpretations are: 1) mean >2.5 means that participants agree with the statements, and 2) mean <2.5 means that participants disagree with the statements.

4 RESULTS AND DISCUSSION

The questionnaire's results from student's and teachers' sides revealed that both of them are seeing DUP as a useful tool in the context of L2 online learning during pandemics. However, students and teachers still faced technical issues. The results details are described below:

4.1 *Students results*

The participants (N = 20) were between the ages of 14 to 15 years. All of them stated that this is the first time they use DUP in online classroom learning during the pandemic. There are 85% (N = 20) of them with no previous experience in learning German and only a few of them 15% (N = 20) have learned the German language. The closed questions results showed that students have positive views of DUP use because all mean scores of all items were >2.5. Meanwhile, the open-ended results showed more depth students' opinion of their experiences using DUP, which supports the results of the closed questions.

Table 1. Students' perception of DUP usage.

Items	Min	Max	Mean
Q1. DUP is a new thing for me and it is interesting.	2	4	3,25
Q2. DUP usage helps me to be more focused on German online learning.	1	4	2,90
Q3. DUPs content encourages me to actively participate in online learning.	2	4	3,10
Q4. DUP helps me to interact and engage with friends in online learning.	2	4	3,10
Q5. DUP helps me to interact and engage with the teachers in online learning.	2	4	3,25
Q6. DUPs content, functions, and extensive toolbars make learning more attractive.	2	4	3,20
Q7. DUPs display and functions make me more enthusiastic and motivated to learn.	2	4	3,10
Q8. DUPs contents promote writing exercises and help to improve my writing skills.	2	4	3,05
Q9. DUPs contents promote listening exercises and help to improve my listening skills	3	4	3,35
Q10. DUPs contents promote reading exercises and help to improve my reading skills	2	4	3,30
Q11. DUPs contents promote speaking exercises and help to improve my speaking skills	2	4	3,10

Item Q1 and Q2 were derived from the second indicator *"deliver learning content: DUP as new improved instructional tools for giving German language learning materials from particular books."* The mean score of Q1 is higher than Q2. This result indicates that DUP successfully attracts the students' attention during online learning. Open-ended answers explained that these were students' first attempts to use DUP in online learning. DUP's colorful display, extensive toolbars, and interactive learning content made it more interesting for students. This result supports the prior research that utilizing an appropriate content delivery tool in online learning is needed because it can determine how students process the content and students' success or failure in learning (Chen et al. 2015). Item Q2 mean score indicates students have a better concentration with DUP in online learning but some of them have not. From open-ended results show that the distractions during online learning with DUP such as the unstable internet connection and it affect the focus-span of students. Teachers need to find an alternate way when they are facing these issues because students need to have a good concentration to shape their volition to learn and use the L2,

Item Q3, Q4, and Q5 were derived from the third indicator *"facilitate interaction – DUP can help to emerge the interactions in the learning environment."* Of the three items, Item Q5 has the highest mean scores among Q3 and Q4. Item Q5 mean score indicates that DUP facilitates the most student-teachers interaction including discussion, asking for clarification, and feedback giving during the online learning. Students-teachers interaction indicates that the learning process is taking place and has a positive influence on the willingness of students to learn L2 (Borup et al. 2012). Prior research showed that the more interaction students had with the instructor, the more they were satisfied with their courses, and the more they thought they learned. Item Q3 mean score indicates that students are more likely to be encouraged to participate in online learning because of DUPs content. Students-learning material interactions lead to satisfactory learning results because students have a chance to involve themselves in some action in online learning (Sun & Chen 2016). Next, item Q4 mean score indicates DUP promotes interaction among students in online learning. Interaction is determined to students' success or failure in language learning, because through interactions, students may reflect on what they have learned, using and connecting prior knowledge and new knowledge, and sharing their experiences, perspectives, and achievements for future learning (Shadiev et al. 2017).

Item Q6 and Q7 were derived from the fourth indicator *"restructure teaching methods – DUPs functions and content help teachers to give more innovative teaching methods."* Item Q6 mean score indicates the learning process is more attractive with DUPs display and various interesting features such as listening and phonetic sections, photos, writing exercises, recording video, and audio. Furthermore, DUP helps teachers to restructure their teaching approaches and brings students comfort and gives them the chance to be innovative during the learning process. Item Q7's mean score indicates that DUP students are more enthusiastic and motivated to learn. This supports the prior findings that digital tool usage can increase students' visual and selective attention that also enhances students' motivation in L2 learning (Stevens & Bavalier 2012). In contrast, prior findings also emphasize that students' motivation in online learning is lower than in the face-to-face learning setting, none less is motivation, the crucial factor of success or failure in language learning (Chen et al. 2015).

Item Q8, Q9, Q10, and Q11 were derived from the first indicators *"technologies promote the practice – DUP promotes four language skills practices."* Among the four questionnaire items, item Q9 has the highest mean score and it means DUPs content facilitates most of German language listening practices for students during online learning. It is because of the easy access and use of audio clips, video clips, and listening exercises that are embedded directly in DUP. Item Q10 mean score indicates that DUP facilitates students to get more exposure and input of various texts that can help them to improve their reading skills. Item Q11 mean scores indicate DUPs content can improve students' speaking skills through audio- and video clips, various speaking practices such as mini dialogue, pictures discussion, and the phonetic section on each section. The least mean score is for item Q8 and indicates that DUPs content helps students improve their writing skills. Various photos, illustrations, and various texts in DUP promote students writing practices and help them to understand how to make their writing tasks. These show that the receptive skills (listening & reading) have a higher mean than productive skills (speaking & writing) and it contrasts with prior studies that showed the productive skills have higher appreciation than receptive skills (Bueno-Alastuey & Pérez 2013). It can be concluded that with DUP students can develop and enhance mostly their listening and reading skills. The possible reasons could be that it is the students' first time using DUP, that has directly embedded audio- and video-clips and interactive texts displayed on it. Otherwise, DUP did not provide adequate assistance to the students in enhancing their speaking and writing skills.

4.2 *Teachers results*

Participants are three female German language teachers, who teach intensive beginner German language classes in one private high school in Jakarta. Teacher A and B have more than five years of teaching experience, and teacher C has been active for one year of teaching German. Teacher A and B, both have stayed in Germany for one year as Au pair Mädchen, whilst teacher C has never stayed in Germany. An Au pair is a single young adult between 18 to 25 of age who has not had

Table 2. Teachers' perception of DUP usage.

Items	Min	Max	Mean
Q1. Using DUP helps support the process of teaching and learning online.	3	4	3,67
Q2. DUPs content helps develop students critical thinking skills.	2	3	2,33
Q3. DUPs content encourages students to participate in online classrooms.	3	4	3,67
Q4. DUPs content encourages students to collaborate with others.	2	4	3,33
Q5. DUPs content encourages students to interact with teachers.	2	3	2,33
Q6. DUPs content, functions, and extensive toolbars make teaching activities interesting.	3	4	3,67
Q7. Using DUP helps to increase students' learning motivation.	3	4	3,67
Q8. DUPs contents promote writing exercises and help students to improve their writing skills.	2	3	2,67
Q9. DUPs contents promote listening exercises and help students to improve their listening skills.	3	4	3,33
Q10. DUPs contents promote reading exercises and help students to improve their reading skills.	2	4	3,33
Q11. DUPs contents promote speaking exercises and help students to improve their speaking skills.	2	3	2,67

children and moves to a foreign country to live with a host family for a specified duration, offering childcare and light housework assistance.

Item Q1 and Q2 were derived from the second indicator *"deliver learning content: DUP as new improved instructional tools for giving German language learning materials from particular books."* Item Q1 mean score is higher than 2.5 and indicates that teachers see DUP as a useful tool in the process of teaching and learning the German language. This result supports the prior findings that teachers' readiness and positive perception toward the instructional tool affect online course satisfaction and learning success (Bhuasiri et al. 2012). In contrast to the answers from the open-ended questions, which show teachers are still not completely understanding or mastering the whole functions of DUP, because of the limited time for teachers to upskill and prepare them to use DUP in response to the COVID-19 pandemic. On other hand, item Q2's mean score is lower than 2.5 and indicates that teachers are unsure that DUP can help students improve their critical thinking skills. Students' critical thinking refers to the thinking procedures that happen in specific situations that depend on the determination of truth, priorities and relevance (Ebadi & Rahimi 2018).

Item Q3, Q4, and Q5 were derived from the third indicator *"facilitate interaction – DUP can help to emerge the interactions in the learning environment."* Item Q3 mean score indicates that DUPs interactive contents can help to increase students' participation during online learning. For instance; group discussion by using DUPs extensive toolbars such as screenshots, interactive white-board, and audio- and video-recording. These support the prior research that stated learner–content interactions affect the objective learning outcomes (Quadir et al. 2019). Item Q4's mean score indicates that teachers agreed DUPs contents help students to collaborate among students during an online classroom. These support the prior finding that among students interaction significantly affects students' achievement on the course (Kurucay & Inan 2017). Next, item Q5's mean score is lower than 2.5 and indicates that teachers disagree that DUP can help them to be more engaged and have more interaction with students. These support Quadir's study (2019) that "learner–teacher interaction did not have a significant influence on objective learning outcomes" (p.11). These contrast with the students' side that claimed they have good interactions with teachers through DUP. However, the amounts of teacher-students' interactions are related to different satisfaction levels and relying on who initiated the interaction and why.

Item Q6 and Q7 were derived from the fourth indicator "*restructure teaching methods – DUPs functions and content help teachers to give more innovative teaching methods.*" Item Q6 mean score indicates teachers believe that DUPs content, function, and extensive toolbars can help them to give more interesting instructions. Interesting means the learning activities considering each student's learning style, and relevant learning materials can attract them to be more focused during the learning process and leads to successful learning. This support for prior studies that claim incorporates "real world" activities in instructional materials enables students to participate more in online learning activities (Cole et al. 2019). Item Q7's mean score indicates teachers agreed that DUP can increase students' motivation because to achieve success in online learning, the appropriate course designs, learning material, and motivation are essential things.

Items Q8, Q9, Q10 and Q11 were derived from the first indicators *"technologies promote the practice – DUP promotes four language skills practices."* Item Q8 mean score indicates teachers agreed that DUPs contents facilitate writing exercise and help students to improve their writing skills. Writing exercises included writing words, sentences, and short text. Item Q9 mean score indicates teachers agreed that DUPs content promotes listening exercises and helps students to improve their listening skills. By using DUP, students can listen to the audio clip directly without needing a DVD to play it. Item Q10 mean score indicates that teachers agreed that DUPs content promotes reading exercises and can help students to improve their reading skills. DUPs various texts are related to the important theme in the real world for students and some of the texts are including illustrations, photos, and the audio clip. Next, Item Q11's mean score indicates teachers agreed that DUPs content facilitates speaking exercises and can help students to improve their speaking skills including pronunciation by using recording tools, also speaking and phonetic sections in each chapter. From these results, it can be concluded that DUP affects more receptive skills than productive skills. This contrast with the prior finding shows that the use of technology in online learning affects more the productive skill than receptive skills (Bueno-Alastuey & Pérez 2013). This difference is depending on the type of instructional tool and learning environment.

In summary, the sudden uses of DUP in online learning during pandemic was seen as a positive action. Because, students and teachers perceived DUP as a useful tool that promotes language skills practices, a medium for delivering learning materials, facilitates interaction during the virtual classroom, and helps teachers particularly in restructuring their teaching methods in online learning.

5 CONCLUSION

This research examined the perceptions of DUP use by teachers and students in online learning during the pandemic. The results show that students and teachers viewed DUP as a beneficial and helpful tool in the process of L2 online learning. These findings supported previous studies that have mentioned both students and teachers often perceive digital tools as interesting tools that can contribute to learning success. DUP can make online learning more interactive and exciting for both parties, and later bring out the interactions in the online classroom. This finding also indicates that DUP is a powerful digital tool for online learning, as the use of DUP reflects the four main benefits of technology in L2 learning, such as Encouraging practices, providing learning materials, facilitating interactions, and restructuring teaching methods. Nevertheless, teachers and students still face some problems, such as technical issues and lack of training sessions for teachers to use DUP. The small number of participants was the limitation of this research. It is strongly recommended to conduct further research with a larger number of participants with different language proficiency levels and focus on how the digital tools can help learners improve their language skills, including listening, speaking, reading and writing skills.

REFERENCES

Bhuasiri, W., Xaymoungkhoun, O., Zo, H., Rho, J. J., & Ciganek, A. P. (2012). Critical success factors for e-learning in developing countries: A comparative analysis between ICT experts and faculty. *Computers & Education, 58*(2), 843–855.

Borup, J., Graham, C. R., & Davies, R. S. (2012). The nature of adolescent learner interaction in a virtual high school setting. *Journal of Computer Assisted Learning, 29*(2), 153–167.

Bueno-Alastuey, M. C., & López Pérez, M. V. (2013). Evaluation of a blended learning language course: students' perceptions of appropriateness for the development of skills and language areas. *Computer Assisted Language Learning, 27*(6), 509–527.

Chen, C.-M., Wang, J.-Y., & Yu, C.-M. (2015). Assessing the attention levels of students by using a novel attention aware system based on brainwave signals. *British Journal of Educational Technology, 48*(2), 348–369.

Cole, A. W., Lennon, L., & Weber, N. L. (2019). Student perceptions of online active learning practices and online learning climate predict online course engagement. *Interactive Learning Environments*, 1–15.

Ebadi, S., & Rahimi, M. (2018). An exploration into the impact of WebQuest-based classrooms on EFL learners' critical thinking and academic writing skills: a mixed-methods study. *Computer Assisted Language Learning, 31*(5–6), 617–651.

Hover, A., & Wise, T. (2020). Exploring ways to create 21st century digital learning experiences. *Education 3-13*, 1–14.

Hromalik, C. D., & Koszalka, T. A. (2018). Self-regulation of the use of digital resources in an online language learning course improves learning outcomes. *Distance Education, 39*(4), 528–547.

Kurucay, M., & Inan, F. A. (2017). Examining the effects of learner-learner interactions on satisfaction and learning in an online undergraduate course. *Computers & Education, 115*, 20–37.

Quadir, B., Yang, J. C., & Chen, N.-S. (2019). The effects of interaction types on learning outcomes in a blog-based interactive learning environment. *Interactive Learning Environments*, 1–14.

Shadiev, R., Hwang, W.-Y., Huang, Y.-M., & Liu, T.-Y. (2017). Facilitating application of language skills in authentic environments with a mobile learning system. *Journal of Computer Assisted Learning, 34*(1), 42–52.

Singh, V., & Thurman, A. (2019). How Many Ways Can We Define Online Learning? A Systematic Literature Review of Definitions of Online Learning (1988-2018). *American Journal of Distance Education, 33*(4), 289–306.

Stevens, C., & Bavelier, D. (2012). The role of selective attention on academic foundations: A cognitive neuroscience perspective. *Developmental Cognitive Neuroscience, 2*, S30–S48.

Sun, A., & Chen, X. (2016). Online Education and Its Effective Practice: A Research Review. *Journal of Information Technology Education: Research, 15*, 157–190.

Thongsri, N., Shen, L., & Bao, Y. (2019). Investigating factors affecting learner's perception toward online learning: evidence from ClassStart application in Thailand. *Behaviour & Information Technology, 38*(12), 1243–1258.

Wei, H.-C., & Chou, C. (2020). Online learning performance and satisfaction: do perceptions and readiness matter? *Distance Education, 41*(1), 48–69.

Wirth & Horn - Informationssysteme GmbH – www.wirth-horn.de. (n.d.). *Digitale Unterrichtspakete von Ernst Klett Sprachen*. Ernst Klett Sprachen GmbH. Retrieved July 2020, from

Zhang, R., & Zou, D. (2020). Types, purposes, and effectiveness of state-of-the-art technologies for second and foreign language learning. *Computer Assisted Language Learning*, 1–47.

Post Pandemic L2 Pedagogy – Adi Putra & Arifah Drajati (Eds)
© 2021 Taylor & Francis Group, London, ISBN 978-1-032-05807-8

A case study of consonant sound problems of Indonesian EFL learners

Diana Ross Arief
Politeknik ATK Yogyakarta (Ministry of Industry), Indonesia

Vinindita Citrayasa
Universitas Atma Jaya Yogyakarta, Indonesia

ABSTRACT: The study investigated the vocational education students' pronunciation problems, especially on English consonant sounds. More specifically, this study looked at 1) what are the students' problems on English consonant sound, 2) what are the dominant problems affecting and 3) how do they resolve their pronunciation problems. The participants were 69 vocational higher education freshmen in Indonesia. The qualitative description approach was used to collect the data. Field observation, questionnaire, and job interview video practice were conducted in the study. The study revealed that most students had problem on the consonants sounds such as: /θ/, /ð/, /dʒ/, /tʃ/, /ʃ/, /j/, etc. Three factors affecting their pronunciation were mother tongue influence, non-corresponding spelling, and age. The most dominant factor was the mother tongue influence. After discussing their partner's feedback, the students said to have a willingness to resolve their difficulties. The results can be used as a reference in making pronunciation instructions to minimize students' pronunciation errors.

Keywords: Consonant sound errors, English learners, pronunciation

1 INTRODUCTION

Mispronouncing words is considered fatal as communication breakdowns will still happen, although students have good grammar and vocabulary command (Celce-Murcia et al. 2010). Research has been conducted not only to investigate the English pronunciation error made by non-native speakers as well as the causal factors but also to seek the most suitable pronunciation instruction to help students coping with the mispronunciation issues as it was in the research on the Jordanian students (Al-Zayed 2017), Hausa speakers (Keshavarz & Khamis 2017) and Sudanese students (Hassan 2014). In conclusion, to acquire proficiency in communication, both students and teachers should collaborate to cope with consonant sound problems to reach proficiency in English communication.

According to (Unubi & Sunday 2019), (Zhang & Yin 2009), and (Zoghbor 2018), various factors were affecting the English pronunciation produced by second language learners (L2 learners). The factors were interference of mother language, learner's age, learner's attitude and psychological, prior pronunciation instruction, and the insufficient knowledge of English phonology and phonetics, including the latest research about the absence of certain English consonant sounds in the student's native language. In summary, pronunciation is essential in communication. Most studies suggested that mother language, learner's age, learner's attitude and psychological, prior pronunciation instruction and the absence of English consonant had become the factors affecting pronunciation. The previous studies mostly focused on the factors. Therefore, more studies should be done to find out how students overcome their consonant sound problems.

Overall, numerous studies focused on teachers to overcome consonant sound problems. Only a few studies measured how students deal with consonant sound problems; because of the reason above, this study was conducted. The solution to fix the students' pronunciation error is to fix the pronunciation instruction. There are numerous teaching pronunciation methods and strategies to

 DOI 10.1201/9781003199267-21

help students get the improvement. Some instruction is focused on the phonetic system, sound feature, articulation, and error correction by using feedback implicitly to raise the students' awareness of their appropriate pronunciation production error. A study related to teaching the sound features was done by Suntornsawet (2019), arguing that failure in pronouncing the initial sound was critical and resulted in a mismatch to the targeted word. Therefore, it is necessary to teach sound features, especially where the sounds are not present in the learners' first language. Besides focusing on the sound feature and articulation, the experimental study used to feature technology such as creating online tasks (Ahmadian & Tavakoli 2010) and using social media networks (Xodabande 2017) to improve students' pronunciation skill mastery. In other words, all methods and strategies to help students improve their pronunciation skills most focused on the teacher instruction while still few studies explored the advantages of students' peer learning strategy in learning and improving their pronunciation. Moreover, Lyster et al. (2013) suggested that the effects, benefits, and strategy of peer learning corrective feedback need to be investigated. Following the facts above, the present study's objectives are to find out the students' problem on English consonant sound problem, the dominant factor, and the way students resolve their problems on English consonant sound problem based on the job interview video.

2 LITERATURE REVIEW

In the EFL and ESL context, numerous studies had been conducted to find out the problematic English sounds as had been done in the research on the Jordanian students (Al-Zayed 2017), Hausa speakers (Keshavarz & Khamis 2017), and Sudanese students (Hassan 2014). Research has been conducted on EFL learners in Iran to find out their error in pronouncing English consonants. The results show that the problematic English consonants are /w/, /v/, /θ/, /ð/ and /ŋ/ and most students have problems pronouncing /v/ (Ercan 2018). While research done by Lee et al. (2019) revealed /a/, /ɑ/, /ɜ/, /o/, /f/, /h/, /v/, /b/, /s/ and /θ/ and /ə/ are problematic for Japanese students.

Several researchers (Ercan 2018; Huang & Jia 2016; Ghorbani 2019; Sardegna et al. 2017) reported the studies regarding the correspondence between the articulatory system and pronunciation mastery. Moreover, the study confirmed the effectiveness of the transcription technique and somehow lend further support to the SLM claim that a target-like perception of the L2 sounds can lead to a target-like production of them and by increasing learners' self-efficacy through optimal challenges and feedback as well as addressing the values of learning pronunciation can be done to achieve pronunciation learning goals. Another study found out that the Japanese students' pronunciation skills improved by associative memory and coding (Saito et al. 2019). In general, most English learners from several countries experienced problematic English sounds (consonants or vowels). Furthermore, to help students overcome the problems, pronunciation instructions such as teaching articulatory systems or transcription techniques might be designed.

The use of technology has benefited students' language production improvement as they have more opportunities to practice. Ahmadian and Tavakoli (2010) suggested that careful online planning and task repetition effectively affect students' accuracy, complexity, and oral production fluency. Xodabande (2017) revealed that using social media networks in teaching English pronunciation is effective and promising so that the technology product will provide more opportunities for students to learn not only inside the class but also outside class. In conclusion, the studies above revealed the use of technology in teaching pronunciation that showed positive results in the effectiveness of methods used to improve students' pronunciation mastery. Therefore, to present a complete work, those studies can be the guidelines to research pronunciation instruction.

3 METHODOLOGY

In this research, the researchers used a qualitative description approach to find out the vocational higher education students' errors based on their job interview video practice in English pronunciation learning. Field observation and questionnaires were conducted to collect the data in the study. This study's participants were 69 (49 females, 20 males) freshmen students of one of the vocational higher institutions in Indonesia, aged between 17–22 years old, majoring in Rubber and Plastic Processing Technology. Most of them had one to ten years' experience learning English, including learning either informal or non-formal institutes. Several of them often practice their English ability through code-switching, joining English language courses, or doing autodidact learning (self-taught).

According to Cohen et al. (2003), a case study is one qualitative approach to investigate and reveal the possible interactions among the participants, events, and other unusual factors. In this study, the field observations were organized during English classroom activities in one semester to collect the data. The observation includes the students' English pronunciation performance, English pronunciation difficulties, and their motivation to overcome their English pronunciation difficulties, especially on consonant sound. The questionnaire had three parts; the first part consisted of the students' educational background. The second part was about students' problems in learning English pronunciation based on their job interview video practice. The third part was composed of five open questions regarding students' opinion on interview practice, the main factor that influenced their English consonant sound problems, the motivation to overcome their difficulties in learning English pronunciation, and suggestion on pronunciation teaching and learning to overcome students' difficulties. The questionnaire results were analyzed by SPSS 26 software.

4 RESULTS AND DISCUSSION

The questionnaire results revealed that most students acknowledged three elements affecting their consonant sound problems on learning English and how they reconcile with the problems. We described the details below:

4.1 *The students' problems on English consonant sounds*

The studies conducted by (Alzinaidi & Abdel Latif 2019), (Anggraini & Istiqamah 2019), and (Nozari & Dell 2012) listed some phonological speech errors consisting of substitution, addition, or deletion of phonemes or, less often, the substitution of entire words, as well as the differences between the writing systems of English and Arabic. The consonant sound errors produced were /v/, /θ/, /ð/, /ʃ/, /ʒ/, and /ŋ/. While based on the job interview recording observation, it was found that most participants in this study had a problem with the consonants sounds production such as strength /θ/, mother /ð/, graduate /dʒ/, question /tʃ/, sure /ʃ/, excuse /j/, etc. The sound of /θ/ was usually replaced by /t/, /ð/ was substituted with /d/, or other sounds in students' L1. Furthermore, Gilakjani and Ahmadi (2011) stated that the students' need to change a conceptual pattern appropriate for their first language in which they have internalized in childhood had become the major problem that second language learners have to deal with pronunciation. These researchers also proposed some factors affecting pronunciation problems such as *accent, stress, intonation, rhythm, motivation and exposure, attitude, instruction, age, personality and mother tongue influence.* In contrast, Zhang and Yin (2009) studied pronunciation problems of English learners in China had defined factors influencing students' English pronunciation problems, such as *interference of Chinese to English, learners' age, learners' attitude and psychological factors, and prior pronunciation instruction.* Finally, the previous studies revealed some factors affecting the students' pronunciation errors.

However, this study mainly discussed the three factors: mother tongue influence, non-corresponding spelling, and age. The questionnaire results revealed that most students acknowledged that those three elements above affected their consonant sound problems in learning English. The open questions portrayed that local language or commonly students' first language (L1) background intervene in producing error consonant sound in learning English. Furthermore, this is in line with what Goswami (2020) pointed out: the more straightforward syllable structure of SHB (Sylheti Bangla) is much simpler than in English; therefore, pronunciation of an English syllable

Table 1. Question 2: The dominant problem affecting the students' English consonant sound.

		Subset for alpha = 0.05	
Difficulties Factors	**N**	1	2
Age	69	2.3043	
Non-corresponding spelling	69		3.1014
Mother Tongue	69		3.1449
Sig.		1.000	.749

Means for groups in homogeneous subsets are displayed.
Uses Harmonic Mean Sample Size = 69.000.

with clusters becomes problematic to the SHB learners of English. While in this study, some English sounds do not exist in students' L1 in which the sounds are unfamiliar for them. Then, Malana (2018) outlined that where the sound is present in respondents' first language, the respondents did not usually commit error; on the contrary, where the sound is absent, the respondents committed the error. Afterwards, limited vocabulary mastery, learners' attitude, and psychological factors such as anxiety, the same sounds with a different distribution, and lack of motivation to learn and practice English pronunciation were remarked as the reasons for their consonant sound difficulty.

4.2 *The dominant problem affecting the students' English consonant sound*

The participants in this study come from various local language backgrounds as their first language such as Sundanese and Javanese from West, Middle, East Java, or Sumatra, Kalimantan, and even Sulawesi. Consequently, it would be much expected that mother tongue or mother language interference had become the most dominant problem faced by the students (3.1449) (demonstrated in Table 1). Most common mispronounced words would be likely caused by the sound of students' first language. It is compatible with Hamidiyah and Arief (2014) who revealed that learners often substitute a particular sound that does not exist in their native language, for example, the sound of /ð/ which occurred 45 times were all the occurrences produced /d/ by the students. Gilakjani and Ahmadi (2011) discussed factors to help teachers distinguish second language learners' problems related to the student's first language interference by designing suitable teaching methods and pronunciation instruction.

Whereas mother language interference was found as the most dominant factor, age was found to be the least or minor problem altering their English consonant sound learning (2.3043). The data is shown in Table 1. One of the fundamental reasons is that the student participants are still categorised as young adult learners who are still likely able to learn the sound system more effectively, while the learning process of adult learners may be more likely inhibited because of their age (Zhang & Yin 2009). In general, because the study participants were only 17–22 years old, age became the least dominant factor for the consonant sound problems, while mother-tongue interference was the most dominant factor regarding the various local language backgrounds.

4.3 *The students' approach to resolving their problems on English consonant sounds*

Despite the students' English pronunciation problems, they have a great willingness to overcome their difficulties. The open question results showed that there are some ways to deal with their problems. Those are; listening to an electronic dictionary to check the standardized pronunciation, checking phonetic symbol in a conventional dictionary, learning from close circle such as ask and practice with their friend or ask their lecturer, do autodidact learning, namely: watching English movie, listening to native speakers' video, listening English song, and learning English in a formal institution. These findings are in correspondence with Magner et al. (2009). They proposed possible solutions to the language barrier. The three categories are informal day-to-day changes in communication patterns, language training, and bridge individuals.

Furthermore, by conducting a job interview video task that involves two students' role play interviewer and interviewee, the students also practice giving and receiving feedback to improve their performance in English pronunciation, grammar and diction. They can observe their peer's performance by playing the videotape and discussing each member's feedback. Most students indicated that the feedback given by their peers made them aware of their problems, mainly when they produced inaccurate consonants of certain words. Afterwards, they checked the correct pronunciation in an online or conventional dictionary and practiced it slowly to master it. The students also realized that listening to English native speakers' spoken language as much as possible may help them to be more familiar with their pronunciation. Because of this reason, some students choose to watch English movies or speech videos, while others prefer to listen to English songs and practice singing. The phenomenon supported Ghaith and Shaaban (2005), in which they stated that Peer Learning strategy is one strategy which can maximize students' learning affection, in this case, involve discussion, self-help study, and team study toward the given feedback as it provides more opportunities for students to be involved in interactive learning. In general, despite the problems faced by students, they were motivated to deal with the problems.

Accordingly, Zhang and Yuan (2020) found that giving explicit pronunciation practice (PI) following presentation-practice-production (PPP) can be beneficial for learners to achieve pronunciation mastery. In contrast, Bui (2016) reported that the teachers' possible mistakes could be acknowledged by the teachers and give proper instruction. Moreover, it is significant to create an environment where students can feel confident and motivated to regularly use their English. Regarding creating a confident and motivated environment, Bosker et al. (2012) suggested that three aspects of fluency (breakdown, speed, and repair fluency) increase the perception of students' fluency. In summary, to support students' motivation to achieve pronunciation mastery, teachers are suggested to design proper pronunciation instruction.

5 CONCLUSION

This study investigated consonant sound problems of English learners in vocational higher education institutions. We presented three research questions to describe: the students' problems on English consonants sounds, the dominant problem affecting the students' English consonant sounds, and how they resolve their problems on English consonants sounds. The research questions were answered in the previous chapter, and the finding summary would be presented as followed. There are four main findings discovered after the data and questionnaire results were analyzed. First, the students' problems on English consonant sound were /θ/, /ð/, /dʒ/, /tʃ/, /ʃ/, and /j/. The error sounds were mainly affected by three factors: mother tongue influence, non-corresponding spelling and age. Second, the dominant problem affecting the students' English consonant sound is mother language interference because of the diversified local languages. Age was found to be the least or minor problem affecting students' English consonant sound like most of the students are aged between 17–22 years old, categorized as young-adult learners. Third, there were some ways for the students to resolve their problems on English pronunciation, namely: listening to the electronic dictionary to check the standardized pronunciation, checking phonetic symbol in the conventional dictionary, learning from close circle such as ask and practice with their friend or ask their lecturer, self-taught English learning, such as: watching English movies, listening to native speaker videos, listening to English songs and learning English in a formal institution. Hopefully, this study will be a reference to design pronunciation instruction to minimize students' consonant errors in English pronunciation. Since the study's findings were too limited to be persuasive, more studies should be conducted on consonant sound problems of English learners in vocational higher education school in a broader range of fields, with a broader range of participants, to provide more complete results. It is anticipated that the results of this study may serve as a reference to conduct more conclusive studies with a greater number of participants.

REFERENCES

Ahmadian, M. J., & Tavakoli, M. (2010).The effects of simultaneous use of careful online planning and task repetition on accuracy, complexity, and fluency in EFL learners' oral production. *Language Teaching Research, 15* (1), 35–39.

Al-Zayed, N. N. (2017). Non-Native pronunciation of English: problems and solutions. *American International Journal of Contemporary Research, 7* (3), 86–90.

Alzinaidi, M.H., & Abdel Latif, M. M. M. (2019). Diagnosing Saudi students' English consonant pronunciation errors. *Arab World English Journal, 10* (4), 180–193.

Bosker, H. R., Pinget, A.-F., Quene, H., Sanders, T., & Jong, N. H. (2012).What makes speech sound fluent? the contributions of pauses, speed, and repairs. *Language Testing, 30*(2), 159–175.

Bui, T. S. (2016). Pronunciations of consonants /ð/ and /θ/ by adult Vietnamese EFL learners. *Indonesian Journal of Applied Linguistics, 6* (1), 125–134.

Celce-Murcia, M., Brinton, D. M., & Goodwin, J. M. (2010). *Teaching pronunciation: a reference for teachers of English to speakers of other languages* (2nd ed). New York, NY: Cambridge University Press.

Cohen, L., Manion, L., & Morrison, K. (2003). *Research Methods in Education.* London, UK: Routledge.

Ercan, H. (2018). Pronunciation problems of Turkish EFL in Northern Cyprus. *International Online Journal of Education and Teaching (IOJET), 5*(4), 877–893.

Ghaith, G. M., & Shaaban, K. A. (2005). Cooperative learning for the disaffected ESL/EFL learners. *The International Journal on School Disaffection, 13* (2), 44–47.

Ghorbani, M. R. (2019). The effect of phonetic transcription on Iranian EFL students' word stress learning. *Journal of Language and Linguistic Studies, 15* (2), 400–410.

Gilakjani, A. P. (2011). A study on the situation of pronunciation instruction in the ESL/EFL classroom. *Journal of Studies in Education, 1*(1), 1–15.

Gilakjani, A. P., & Ahmadi M. R. (2011). Why is pronunciation so difficult to learn? *English Language Teaching, 4*(3), 74–83.

Goswami, A. (2020). Changing contours: the interference of the mother tongue on English speaking Sylheti Bengali. *Journal of English as an International Language, 15*(1), 100–134.

Hamidiyah, A. & Arief. D. R. (2014). Problematic English consonant sounds production of EFL students in Banten. Proceeding at The Eleventh International Conference on English Studies.

Hassan, E. M. I. (2014), Pronunciation problems: a case study of English language students at Sudan University of Science and Technology. *English Language and Literature Studies, 4*(4), 31–44.

Huang, X., & Jia, X. (2016). Corrective Feedback on Pronunciation: Students' and Teachers'. *International Journal of English Linguistics, 6*(6), 245–254.

Keshavarz, M. H., & Khamis Abubakar, M. (2017). An investigation into pronunciation problems of Hausa speaking learners of English. *International Online Journal of Education and Teaching (IOJET), 4*(1), 61–72.

Lee, B., Plonsky, L. & Saito, K. (2019). The effects of perception- vs. production-based pronunciation instruction, *System, 88*, 1–35.

Lyster, R., Saito, K., & Sato, M. (2012). Oral corrective feedback in a second language classroom. *Language Teaching, 46*(1), 1–40.

Magner, U., Kathrin, K., & Anne-Will, H. (2009). Babel in business: the language barrier and its solutions in the hq-subsidiary relationship. *Journal of World Business, 46*(3), 279–287.

Malana, M. F. (2018). First language interference in learning the English language. *Journal of English as an International Language, 13*(2.2), 32–46.

Nozari, N., & Dell, G. S. (2012). Feature migration in time: Reflection of selective attention on speech errors. *Journal of Experimental Psychology: Learning, Memory, and Cognition, 38*(4), 1084–1090.

Saito, K., Suzukida, Y., & Sun, H. (2019). Aptitude, experience, and second language pronunciation proficiency development in classroom settings: A longitudinal study. *Studies in Second Language Acquisition, 41*(1), 201–225.

Sardegna, V.G., Lee, J., & Kuseyc, C. (2017). Learning self-efficacy, attitudes, and choice of strategies for English pronunciation. *Language Learning, 68*(1), 1–32.

Suntornsawet, J. (2019). Problematic phonological features of foreign-accented English pronunciation as threats to international intelligibility: Thai EIL pronunciation core. *Journal of English as an International Language, 14*(2), 72–93.

Unubi & Sunday, A. (2019). A contrastive study of English and Igala segmental phonemes: implications for ESL teachers and learners. *Journal of Biomedical Engineering and Medical Imaging, 6*(6), 36–43.

Xodabande, I. (2017). The effectiveness of social media network Telegram in teaching English language pronunciation to Iranian EFL learners. *Cogent Education, 4*(1), 1–14.

Zhang, F. & Yin. P. (2009). A study of pronunciation problems of English learners in China. *Asian Journal Science*, 5 (6), 141–146.
Zhang, R., & Yuan, Z. M. (2020). Examining the effects of explicit pronunciation instruction on the development of L2 pronunciation. *Studies in Second Language Acquisition, 42*(4), 905–918.
Zoghbor, W. (2018). Teaching English Pronunciation to Multi-Dialect First Language Learners: The revival of the Lingua Franca Core (LFC). *System, 78*, 1–14.

Post Pandemic L2 Pedagogy – Adi Putra & Arifah Drajati (Eds)
© 2021 Taylor & Francis Group, London, ISBN 978-1-032-05807-8

Brainstorming-based project learning in a German reading classroom

Dian Permatasari Kusuma Dayu & Liya Atika Anggrasari
PGRI Madiun University, Indonesia

Nur Handayani
Senior High School 1 Prambanan, Indonesia

ABSTRACT: This study aimed to determine the effect of the implementation of brainstorming-based project learning models, specifically on the use of brainstorming, in a German reading classroom. This research used an experimental research method. The research design used a pre-experimental one-group pretest-posttest design. Data analysis was conducted by using a quantitative design with a t-test. Data of the research included the data of pretest scores and post-test scores in the experimental and control classes. This result showed that the T-arithmetic > T-table or 4,643 > 1,801 at the significant level 0.05. Brainstorming-based project learning model was proven successful in improving the German reading skills of the students. The problem occurring in German learning was reading skills because students thought that German was a new language and felt difficult to understand its grammar. Besides, students' difficulty in learning and reading German was their lack of vocabulary that made it difficult to understand the reading text. This study also provided insights for the teachers to implement brainstorming-based project learning in their classrooms. Further research was needed to explore more about students' reading activities on learning German in which we would like to discuss further within.

Keywords: Brainstorming, German language classroom, L2, project based learning, reading

1 INTRODUCTION

Learning German in reading skills is closely related to text comprehension skills. Students often face difficulties in learning and in reading German, for example, in understanding the texts, such as determining the theme or title of a text and answering questions related to the text being read. The difficulties experienced by these students also affect when they express ideas or thoughts, both in oral and in written form. The challenges found and the difficulties encountered in learning German include the aspects of language, such as vocabulary and grammar. Lack of mastery of vocabulary and grammar can be seen when students often ask the teacher to interpret each German word while learning. Students also still have difficulty understanding and finding the main idea of the text to be read.

In the last ten years, there have also been several studies investigating the challenges that students and teachers faced in L2 classrooms. In a study by Hussain (2017) a writing and reading activity is a complex process that involves acquiring various skills that contribute to overall writing difficulties for any language user. The teacher still uses the discussion or lecture learning method which causes learning to read seem less attractive, and students think that learning to read is a boring activity (Jaya 2018). Besides, they also have difficulties understanding the text. When the students read a text, they cannot understand and explain the meaning of the text well. They are difficult in answering questions. In reading German text, the teacher uses the translation method and reads German text (Hidayanti et al. 2018). In the process of learning, the teacher translates every word in the text, but it gives an impact on students' lack of reading skills. They cannot understand a text in German easily without any help from the teacher. Moreover, they also do not have an interest in reading a text. It

DOI 10.1201/9781003199267-22

can happen because of inaccurate learning strategies or methods used by the teacher when learning to read (Yuan 2010; Hidayanti et al. 2018; Dahlberg 2016). In general, all of these studies discuss problems in learning to write in foreign languages. The discussion above proves that the students can overcome difficulties in understanding the German language especially in reading activity by using several active learning models and making this learning easier for students to write and read.

The brainstorming learning model based on project learning as a strategy in the written process has not been implemented to write German. So far, only the researchers use strategies outside of brainstorming-based project learning so that the brainstorming-based Project learning model needs to be used to teach reading in German. Based on this problem, the researcher argues that there is a need to improve the learning process with an attractive method in order to help the students to take an active role and be more interested in learning German. This study aims to determine the effectiveness of brainstorming-based project learning models in reading skills. In the learning process by using the Brainstorming-based project learning model, it is expected to improve German reading skills. Besides, writing and reading is the ability not only to get ideas from thoughts onto paper but also to generate more meaning and make ideas clear.

2 LITERATURE REVIEW

The implementation of the brainstorming model in the learning process as a strategy in the reading activity can overcome the students' learning obstacles in understanding German reading. Brainstorming-based project learning has the advantage of being able to process learning activities. The advantage is that the participants have an enthusiasm to build their creative ideas (Michinov & Primois 2005). The brainstorming-based project learning model can be a learning model that facilitates creative thinking in someone, and this model is more effective than the group model (Brown & Paulus 2002). Project-based learning is an approach to the learning process that can improve the students' thinking power in schools and colleges because learning activities focus on the students' outcomes to complete projects (Karaçalli & Korur 2014).

Brainstorming techniques are easy techniques that can encourage success in writing and reading. Often students have difficulty in understanding the text and expressing ideas that match the theme, or even students do not know what to say (Brown 2001; Aldeirre et al. 2018; Philips 2008). The Brainstorming model is a technique to provoke and encourage students' creative thinking based on general ideas in group discussions (Weichbroth 2016). Project-based learning has maintained a wide range of learning outcomes, including conceptual knowledge, basic skills, and motivation (Darling-Hammond 2008; Guo et al. 2020). These matters can influence students in a more active and effective learning process than uses conventional learning models in project-based learning. They can use project-based learning in literacy, mathematics, and social studies lessons. (Jessica & Mark 2020; Kingston 2018). In conclusion, the above studies show positive results in the effectiveness of the brainstorming learning model and project-based learning in the learning process. This learning model is used to improve reading skills.

Brainstorming-based project learning models is a learning model that aims to train students to understand based on their understanding and to interpret a reading by discussing project guidance done in groups. In its implementation, learning to read German can be understood by teachers and students. There are 5 steps in the brainstorming-based project learning model is a learning model namely; (1) The teacher delivers the learning material, (2) The teacher divides groups of 3–5 people, (3) The teacher shares the discussion theme, and the students create a class project discussion, (4) the teacher monitors the project of discussions, and (5) the teacher holds an evaluation. Using a brainstorming-based project learning model can make it easier for students to understand the reading activity. The research results have proven that this research is good to use as a method to help students remember what they read and can help the teaching and learning process in class, which is carried out by reading German texts. In the implementation process, reading learning activities using the brainstorming-based project learning model gives students the freedom to understand the meaning of reading. Also, in the learning process, students make a reading summary made in group discussions.

3 RESEARCH METHODOLOGY

To collect the data, field observation and tests were conducted in the study. Participants in this study were 60 (40 female, 20 male) male and female students from secondary schools in Indonesia aged 15–17 years. We recruited them with a simple random sampling method. Simple random sampling technique or simple random sampling to determine the treated class and the untreated class.

This quantitative study used a pre-experimental-one-group pretest-posttest research design. A one-group pretest-posttest design is a type of research design most often utilized by behavioral researchers to determine the effect of a treatment or intervention on a given sample (Knapp 2016). The study used questionnaires and tests. The instrument was used in a reading test with a multiple-choice test consisting of long passages and short readings. The data analysis technique used a Quantitative design to see the difference in reading skills between the experimental class and the control class was a t-test. The data in this study included data on pretest scores and posttest scores in the experimental and control classes to see the differences in the students' reading skills between the implementation of brainstorming-based project learning Model with the implementation of conventional learning models. The statistical analysis was aimed to find out normality distribution, Homogeneity of variance, and effectiveness test.

4 FINDINGS AND DISCUSSION

The results show that by using the Brainstorming-based project learning model, German reading skills increased. The details are described below:

4.1 *Data posttest score data of reading ability in the control class*

Table 1. The analysis result data posttest score data of reading ability in the control class.

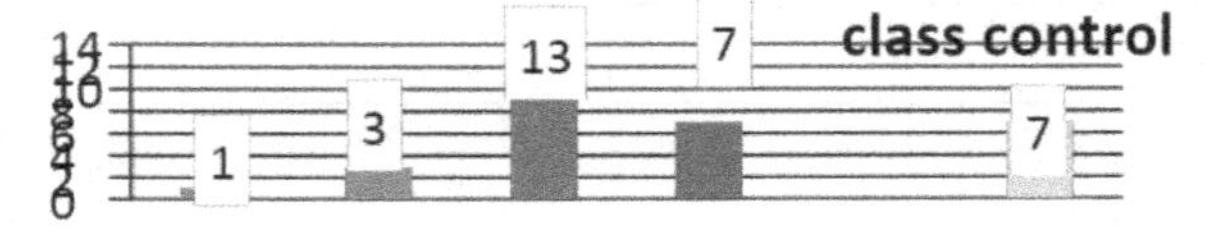

The experimental class is a class that gets a learning treatment without using a Brainstorming-based project learning model on reading skills. Data was based on the results of the statistical analysis of SPSS for Windows 13.0, the overall minimum value is 21, the maximum value is 29, the average score (mean) is 29.87; mode of 28.00; the median of 28.00; and the standard deviation (standard deviation) of 2.013. We can see the frequency of distribution of German reading skills in this group's data in the following table.

4.2 *Data posttest score data of reading ability in the experiment class*

Table 2. The analysis result data posttest score data of reading ability in the experiment class.

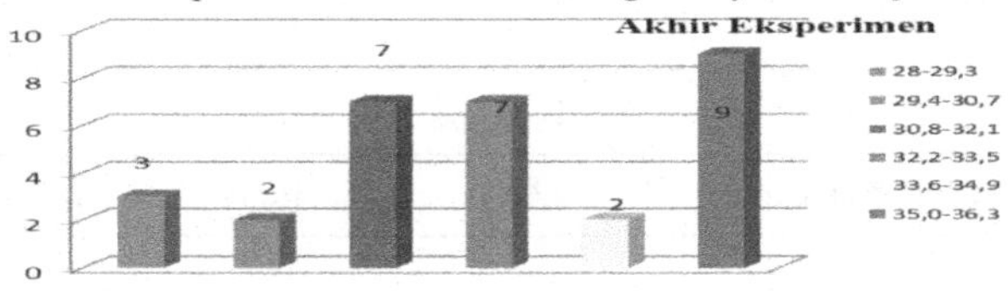

The experimental class is a class that gets a learning treatment with a Brainstorming-based project learning Model. Data Based on the results of the statistical analysis of SPSS for windows 13.0, the overall minimum score is 28, the maximum value is 36, the average score (mean) is

31.87; mode amount of 32.00; a median of 32.00; and the standard deviation (standard deviation) of 2.013. We can see the frequency of distribution of German reading skills in this group's data in the following table.

4.3 *The normality test*

The normality test is also carried out on the data distribution of the German reading skills scores in the control class and the experimental class. The distribution of normal effective value data is due to the normality test result of the experimental class is 0.175, and the control class is 0.321. The normality test is greater than 0.05.

Table 3. Normality test.

Class	Asymp. sig. (2-tailed)	Conclusion
Control	0,321	Normal distribution
Experiment	0,175	Normal distribution

The table above shows the results of the gain score calculation, which have normal data distribution. This is because the gain score has a value (asymp. sig) of more than 0.05, namely 0.325 in the control class and 0.175 in the experimental class.

4.4 *The homogeneity test*

After the data distribution normality test is carried out, the homogeneity test is then carried out. The condition is homogeneous if the calculated significance value is greater than the significance level of 0.05 (5%). The results of the homogeneity test can be seen in the following table:

Table 4. Homogeneity test.

Asymp. sig. (2-tailed)	Conclusion
0,242	Sig.0,242 > 0,05 = Homogen

From the table above, we see that the pretest data sig value of the experimental class and control class is 0.242. It shows a sig value of more than 0.05. Thus, it also obtained the student's reading skills decisions from the homogeneous pretest data. It is stated to be homogeneous because the significance value is greater than the significance level.

The T-test aims to test the hypothesis in order to determine the differences between both the groups that take learning to read using the brainstorming-based project learning model and the groups that take learning to read German without using the brainstorming-based Project learning model. The T-test is also used to test the effectiveness of using the brainstorming-based project learning model in reading learning compared to reading learning without using the brainstorming-based project learning model.

Table 5. T-test.

Treatment	$T_{arithmetic}$	T_{table}	Conclusion
The brainstorming-based project learning	4,643	1,801	H_0 was rejected/H_1 accepted

Based on the t-test result, there is a difference between the experimental class using the brainstorming-based project learning model and the control class using the discussion learning model in the hypothesis using the T-test. The data analysis result of the brainstorming-based project learning model toward German reading skills gets T-arithmetic = 4.643 and T-table = 1.801. According to the criteria test, this research concludes that Ho is rejected or H1 is accepted since the result shows T-arithmetic > T-table or 4.643 > 1.801. In other words, there is a significant influence of the implementation of the Brainstorming-based project learning model toward the students' improvement in the German reading skill.

Brainstorming-based project learning model is one of the learning models used by teachers in class. In the learning process, the teacher provides reading texts to students, discusses them in groups, and answers or expresses opinions or comments toward the text. Then in the last session, students discuss together to make the final product on the text read. The students' reading skills using the brainstorming learning model are better than reading skills using conventional learning models. The use of a brainstorming-based project learning model also makes it the students easier to learn German text because they can express their ideas and opinions freely. If they get some difficulties, they are welcome to ask the teacher. The use of the brainstorming-based project learning model can create a friendly atmosphere between the students and teachers.

This condition makes the students' interested and motivated to learn German, especially learning reading skills increase. A conducive learning situation can support them to master the material presented by the teacher, resulting in good reading skills and significantly better results than those taught by conventional techniques. Project-Based Learning is a way of learning that leads to a training process based on real problems done by yourself through certain activities (projects) (Gijbels 2005). Project-Based Learning is a learning method that encourages students to apply critical thinking and problem-solving skills and gain knowledge about the real problems and issues it faces. In this project-based learning, educators will play a more role as facilitators who guide students through the learning process. Planning and implementing project activities is time-consuming and time-consuming activity. Its implementation also requires careful consideration (Habok & Nagy 2016; Rogers et al. 2011).

The learning atmosphere with brainstorming-based project learning models is very different from learning with conventional techniques. The students are more active and enthusiastic because of the social interactions in group work to convey their ideas freely. The learning center is student-center while the teacher only acts as a facilitator to make the students learn independently. Learning with the Brainstorming learning model based on project-based learning is fun and not boring because the students can work together to answer questions from the teacher and have their respective roles in the group. Baker and Westrup (2005) stated that when students have fun learning, they will take a more active part in the lesson with a variety of interesting and guided learning. Brainstorming-based can be seen as a technique where individuals or groups engage in critical thinking to produce broad ideas and creative solutions to a problem.

Brainstorming can be effective and becomes important to implement it with an open mind and a spirit of non-judgment. In other words, if the students feel fun, enjoyable, and comfortable during the teaching-learning process, they can catch the maximum comprehension of the lesson material. Teachers' implementation of Project-based learning has greatly affected students' content understanding and development of skills (Kokotsaki et al. 2016). Brainstorming is one technique for fostering group creativity by which ideas and thoughts are shared among members spontaneously to reach solutions to practical problems (Gogus 2012; Hosam 2017). The research above shows that the brainstorming-based project learning model can make it easier for students to understand German reading texts.

5 CONCLUSION

At the conclusion of this study, the brainstorming-based project learning model is effective for reading German. Implementing the brainstorming-based project learning model can improve students' reading skills in German. The brainstorming-based project learning model can motivate students, train students to work together to complete projects, and make it easy for students to understand

German reading given by the teacher. The study of brainstorming-based project learning models got the students' data result in their German reading skills. Students are also very enthusiastic about getting the material needed and can follow the teaching and learning process well. Students need to be trained in a guided manner to stand a German text in learning reading skills. Therefore, teachers should be able to use a learning model that can improve students' reading skills. In further research, the brainstorming-based project learning model expects to facilitate the learning process of Indonesian reading skills and make it easier for students to express their creative ideas in story writing activities. Based on the conclusion, brainstorming-based project learning can apply as an alternative in learning to read German to help students in critical thinking skills and understand the meaning of reading in German.

REFERENCES

Al-Samarraie, H., & Hurmuzan, S. (2017). A review of brainstorming techniques in higher education. *Thinking Skills and Creativity, 27*, 78–91.

Brown, H. G. (2001). Teaching by principles: Interactive approach to language pedagogy. New York, NY: Longman.

Brown, V. R., & Paulus, P. B. (2002). Making Group Brainstorming More Effective: Recommendations From an Associative Memory Perspective. *Current Directions in Psychological Science*, 11(6), 208–212.

Baker & Westrup. (2005). Practical English language teaching. New York, NY: McGraw-Hill.

Dahlberg, L. (2016). Learning strategies for reading and listening in the Swedish national syllabus for English: a case study of four English language teachers' best practices.

Darling-Hammond, L. (2008). Teaching and learning for understanding. Powerful learning: What we know about teaching for understanding. San Francisco, CA: Jossey Bass.

Gijbels, D. (2005). Effects of Problem-Based Learning: A Meta-Analysis From the Angle of Assessment. Educational Journal Research. 75,1.

Gogus, A. (2012). Brainstorming and Learning. *Encyclopedia of the Sciences of Learning*. Boston, MA: Springer US.

Guo, P., Saab, N., Post, L. S., & Admiraal, W. (2020). A review of project-based learning in higher education: Student outcomes and measures. *International Journal of Educational Research, 102*, 101586.

Hidayanti, W., Rochintaniawati, D & Agustin, R. (2018). The Effect of Brainstorming on Students' Creative Thinking Skill in Learning Nutrition. Journal of Science Learning. 1, 44.

Habók, A., Nagy, J. (2016). In-service teachers' perceptions of project-based learning. *Springer Plus* **5,** 83.

Hussain, S.S. (2017). Teaching Writing to Second Language Learners: Bench-marking Strategies for Classroom. *Arab World English Journal, 8* (2).

Jessica, D. C & Mark W. (2020) Hands-Joined Learning as a Framework for Personalizing Project-Based Learning in a Middle Grades Classroom: An Exploratory Study. *Research in Middle Level Education,* 43:2, 1–17.

Karaçalli, S. and Korur, F. (2014), The Effects of Project-Based Learning. *School science and mathematics*, 114: 224–235.

Kingston, S. (2018). Project-based learning & student achievement: What does the research tell us. *PBL Evidence Matters*,1(1), 1–11.

Knapp, T. R. (2016). Why Is the One-Group Pretest–Posttest Design Still Used?. *Clinical Nursing Research, 25*(5), 467–472.

Kokotsaki, D., Menzies, V., & Wiggins, A. (2016). Project-based learning: A review of the literature. *Improving Schools*, 19(3), 267–277.

Naser. A. (2015). The Effect of Using Brainstorming Strategy in Developing Creative Problem Solving Skills among male Students in Kuwait: A Field Study on Saud Al-Kharji School in Kuwait City. *Journal of Education and Practice.* 6(3).

N. Michinov, C. Primois. (2005). Improving Productivity and Creativity in Online Groups through Social Comparison Process: New Evidence for Asynchronous Electronic Brainstorming. *Computers in Human Behavior*. Vol. 21, No. 1, pp. 11–2.

Rogers, M.A.P., Cross, D.I., & Gresalfi, M.S. (2011). First Year Implementation Of A Project-Based Learning Approach: The Need For Addressing Teachers' Orientations In The Era Of Reform. *International Journal of Science and Mathematics Education*. **9,** 893–917.

Post Pandemic L2 Pedagogy – Adi Putra & Arifah Drajati (Eds)
© 2021 Taylor & Francis Group, London, ISBN 978-1-032-05807-8

Instagram: Digital platform for promoting ELLs' multimodal literacy in narrative writing under TPACK-21CL

Sri Haryati
English Department, Universitas Sebelas Maret, Indonesia

Eka Nurhidayat
English Education Study Program, Universitas Majalengka, Indonesia

Itok Dwi Kurniawan
Department of Procedural Law, Universitas Sebelas Maret, Indonesia

ABSTRACT: Multimodal literacy is an essential literacy skill for university students. It is required for multimodality to interpret the semiotic mode in making meaning. The representation of semiotic modes such as color, visual, graphic, gestures, spatial, etc., provides input for multimodal literacy development. Regarding the development of technology in ELT, the integration of technology shifts the paradigm that multimodal text is better than printed text. Thus, the involvement of digital multimodal text in social media like Instagram is undeniable. Digital images and video and text as the excellent features of Instagram provide a chance for developing students' multimodal literacy in writing narrative text. The wealth of semiotic resources is presented in the digital images that provide an excellent narrative writing source. In creating the multimodal text being posted on Instagram and to critically view multimodal texts, students need to understand how meanings are generated from modes and be able to identify textual evidence to support the understanding of multimodality in the text.

Keywords: digital photograph, multimodal literacy, semiotic modes, TPACK-21CL

1 INTRODUCTION

In the 21st century, the understanding of literacy is different from previous times. At the cost of other representational forms of sense formation, literacy was primarily related to language. However, radical advancement in the field of technology and communication redefined the communication landscape, broadened the boundaries of literacy, and gave rise to new types of texts in which the way of making meanings was not based solely on language but on a multimodal way of making meanings (Jewitt in Loerts and Rachel Heydon 2016; Papadopoulou et al. 2019). The understanding shift of literacy breaks the traditional literacy that restricted the ability for writing and reading to the meaning-making process based on the presented semiotic modes. The use of many semiotic resources or modes in the design of a semiotic product or case, along with the particular way these modes are combined to give meaning (Kress and Van Leeuwen in Papadopoulou et al. 2019).

The semiotic resources or modes significantly contribute to meaning-making; thus, literacy on multimodalities is required. Multimodal literacy is the ability to encode or decode language, visual, spatial, gestural, and audio meaning modes within texts that alter the meanings of words in different digital and cultural contexts (Pahl and Rowsell in Mills et al. 2016). Multimodality, defined as the active and dynamic interrelationship between the various semiotic modes (linguistic, visual, gestural, spatial, or audio mode) which individuals may use to extract meaning during text interaction because writing and reading paper-based texts are not enough to communicate

DOI 10.1201/9781003199267-23

through multiple-meaning platforms (Bezemer and O'Halloran 2016; Kress 2010; Papadopoulou et al. 2019). For the development of critical literacy, which is important for student education, multimodal literacy is therefore required, as contemporary texts are multimodal and language-centered approaches are not adequate for the process of meaning in the text.

Regarding the development of technology in 21CL, It is generally agreed that the notion of literacy should be broadened as digital texts proliferate (Tan et al. 2019). The technology movement from printed text to digital text thus stimulates mobile applications and social media as tools for academic purposes, generates activities for them online, or keeps in touch with students (Zhang 2013). Instagram as one of the trending social media is considered as the education platform to promote ELLs' multimodal literacy in digital narrative writing. Instagram as a photography-based smartphone application provides a semiotic mode and varying modalities (Borges-Rey 2015; Kylie Budge 2020) which can promote the students' multimodal literacy. Current multimodal and digital platforms such as Instagram highlights the need to understand teachers' TPACK (technological pedagogical and content knowledge) in 21CL for reframing literacy to better respond to the multimodal communication landscapes. Empirical research on TPACK in language education also affirms the significance of TPACK in the effective implementation of mobile-assisted language learning (Tan et al. 2019).

Considering the essential of multimodal literacy for ELLs and the excellence of digital photography (images) and video presented in the trending social media, Instagram, this article discusses the use of multimodal text in form of digital photography and video posted in Instagram to develop ELLs' multimodal literacy in narrative writing. Digital photography (image) and video, categorized as multimodal text, is the source of semiotic modes as the basic resource in meaning-making. Narrative writing is presented based on the multimodal text posted on Instagram. Instead of developing ELLs' multimodal literacy in narrative writing, the use of digital photography (image) and video is assumed to increase the students' engagement in digital narrative writing.

2 SOCIAL SEMIOTIC AND MULTIMODAL LITERACY

Across several different ways, the idea of multimodal literacy has been discussed, including classroom research with literature and second language learning (Unsworth and Royce in El Refaie 2010). The term 'multimodal literacy' means the ability to use sensory modalities (linguistic, gestural, audiovisual and spatial) to generate and understand semiotic resources or modes (language, image, and gesture) to make meaning (Rhoades et al. 2015; Mills & Katherine Doyle 2019). Significance-making happens at various levels through reading, viewing, understanding, reacting to, creating, and communicating with multimodal texts and multimodal communication. Listening, speaking and writing, designing and producing such texts may be involved (Walsh in Friedman 2018).

Modes are socially and culturally developed instruments from a social semiotic perspective to establish a context that is usually interrelated. Therefore, the derivation of meaning in texts is based on all the semiotic modes available, which are often used in conjunction (Mills & Unsworth 2017, p. 5). However, since textual meaning is not constructed in the abstract but in particular contexts, multimodality can be used within a particular context and culture to improve the interpretation of the meaning of communication occurrences, documents, and objects (O'Halloran in Papadopoulou et al. 2019).

Two or more semiotic forms, textual, visual, auditory, gestural and spatial make up multimodal texts. These modes are semiotic instruments with their own unique meaning-making process for each mode (Tan et al. 2020). Through explicit teaching of this metal language, its emphasis is on the analysis of images, i.e. the visual mode, while the visual mode is often used as a basis to analyze other semiotic modes that contribute to the meaning-making process. Multimodality relating to visual communication and the interaction of verbal and visual modes is significant in understanding and producing multimodal texts. Multimodality derives from the theory of social semiotics, which examines how individuals use semiotic methods to establish meanings in various social and cultural practices (Kress 2015). From this point of view, definitions are pluralistic, and defining the semiotic capacity of a communication mode and challenging the meanings of each mode is a central inquiry in social semiotics. Therefore, texts are better interpreted as motivated signs focused on the desires and intentions of the sign maker in relation to the social semantic perspective (Kress 2015). The structure of a text points to the variety of social and cultural circumstances that affect the sign-makers in the contexts in which they are situated and their interactions with the target audiences. Critical reading allows the reader to degrade the deep meaning of any event, document, technique, method, entity, statement, picture, or situation when the definition of text is understood from a social semantic point of view (Lankshear, in Tan et al. 2019).

3 DEVELOPING ELLS' MULTIMODAL LITERACY IN TPACK-21CL

The rapid advancement of technologies such as smartphones, laptops, and tablets makes it easy to access a wide variety of semiotic opportunities in the form of multimodal text. In the field of education, educators currently understand that other influential learning opportunities are created by multimodal forms made increasingly possible by digital technology (Li 2017; Manderino & Castek 2016; Jones et al. 2020). The idea that 'post typographic multimodal texts consist of written language and typically convey meaning through other visual components such as script, typography, color, and illustration (Refaie & Hörschelmann 2010). It affects literacy activities in which it is assumed that images are better able than words to express relationships and degree problems (Lemke 2002) and to evoke an immediate emotional response (words tend to be more pertinent to the role of delivering causality and temporal sequence). The use of photography (images) has also been gathering momentum for pedagogical needs. It is considered a pleasant and attractive educational medium, which encourages the active involvement of verbal expression (Friedman 2018).

Compared to the past teaching where the printed text is merely used as the source of learning, the multimodal text is assumed to better reinforce the students' learning. A lot of studies have started to pay attention to the multimodality of communication in this modern world with image analysis, image and text interplay, and multimodality analysis in education. It is undeniably true that the screen has profoundly affected meaning formation, and the relationship between text and picture has changed drastically. (Pietikäinen and Pitkänen-Huhta 2013). Two or more semiotic modes, namely the linguistic, visual, audio, gestural, and spatial modes, consist of multimodal texts. With each mode having its own special meaning-making method, these modes are semiotic resources (Tan & Zammit 2018). The interaction between the multimodal text and the audience in the multimodal text is focused on questions such as how to engage the artifact (through sound, colour, humor, picture choice), how to make key facts/concepts salient (color, size or weight, repetition, focus) and how to encourage the comprehension of learners through 'abundance of messages' or the use of multiple meaning-making tools (Hammond and Gibbons in Jones et al. 2020).

Research has shown that learners prefer multimodal texts rather than printed texts (Bearne et al. 2012; Gecer and Dag 2012; Tuzel 2012, 2013). The use of multimodal texts leads to changes in the method of teaching, assessment and evaluation in the classroom, class engagement, and teacher and learner responsibilities (Walsh and Munns et al. in Tuzel, 2013). Multimodal texts enable one to respond to and manage each other's listening, seeing, and reading in the process of generating meaning (Haggerty in Tuzel, 2013). Students report more thoroughly using their cognitive skills, more creativity and being more informed, participatory and productive in the classroom using multimodal texts in classrooms (Callow and Zammit 2012; Lin et al. 2013).

4 INSTAGRAM FOR PROMOTING MULTIMODAL LITERACY IN NARRATIVE WRITING

Promoting ELLs' multimodal literacies is closely related to the pedagogical changes that need teachers' technology sensitivity. Since as part of TPACK, multimodal literacies appear to be correlated with the use of ICT, the element of technical and pedagogical knowledge (TPK) may also need to include cyber wellness products. In addition, the Social Semiotic (KS) element now reflects the core efficacy of the awareness of teachers' content knowledge (CK). This factor will need to be extended to differentiate between metal language products for written texts and multimodal texts. (Tan et al. 2019). Considering the role of the TPACK framework related to promoting ELLs' multimodal literacy, the technology integration in ELT should be based on the representation of multimodal text to stimulate learning. Teachers' can make use of the digital platform where the multimodal text is presented or created as the learning result. Instagram, the trending social media, is one of the alternative digital platforms as the wealth of multimodal text in the form of digital photography, images, and video are accommodated there.

The social photography trend in the twenty-first century is Instagram. It is the "shoot and share" platform where a lot of social life is represented, explored and then shared with the globe through vast algorithm-enabled social networks and the active use of those who interact with the platform. Launched in 2010 as a smartphone application based on photography, Instagram has gained popularity since April 2012. (Ferrara et al. 2014). Instagram's technology allows for both still photography and video (Budge (2020). Launching photo-sharing as a photo snap, such as video, text and story sharing, has been added to significantly contribute to its development (Ellison 2017). In terms of language learning, Instagram has four language skills in and outside the classroom to practice the language. In addition, some studies have been carried out on Instagram to develop writing skills (Soviyah & Etikaningsih 2018). Instagram, the world's most popular social media site (Smith & Anderson 2018), is a social media platform where users can share self-generated content in the form of images or videos (Abney et al. 2018).

Images are useful because they are used as resources to explain patterns and processes that cannot be clearly observed and relationships that are difficult to define orally (Meneses et al. 2018). Images, as well as videos in Instagram posting, contain semiotic resources or modes as the core elements of the multimodal text. Using the power of modes in digital images, photography, and videos are expected to develop ELLs' multimodal literacy in producing Narrative Writing. According to the Alabama Department of Archives and History in Pérez-Gómez and Daza (2019), narrative writing consists of recounting a series of acts that occur in a given time as conveniently as possible to a reader who is supposed to understand the events and to observe the sequence of events conveyed in the storytelling. As Character-Problem-Solution Narrative, there are two patterns of narrative writing that revolve around a problem or difficulty that the main character is facing and seeks to find the most probable solution and personal experience narrative where the character tells the story in the first person and talks of an attractive and attention-grabbing experience, offering explanations and information.

Writing a story or narrative writing is a part of literacy where the master of multimodal literacy presented in Instagram posts helps ELLs better produce a text. The representation of semiotic resources such as linguistic, color, visual, audio, gestural, and spatial in digital photography and video is assumed to stimulate multimodal literacy. With the semiotic resources' help, the students are also engaged and motivated to create the multimodal narrative text. The narrative writing in the Instagram post is presented in Figure 1 and Figure 2.

The narrative writing presented in the Instagram post is based on the principle of multimodal text that for presenting events in sequence, writing is easier. The picture is best for reflecting the relationship between elements in space. (Kress in Lim, 2018). The sample of narrative writing presented in Figure 1 is a kind of personal experience narrative where the author told her personal story about the authors' mother and her experience with her. The authors' emotions and expressions flow out of her mind into a past story. Besides the text, the Insta-post is completed with digital photography as a multimodal text. The authors' mother's portrait in black and white mode contains multimodal modes that lots of interpretation can be drawn. The figure of the authors' mom that is represented in the picture was also written in narrative writing.

Using the same model of narrative writing an Instagram post, the narrative writing teaching that sharpening the students multimodal literacy through the representation of digital photography for university students can be conducted under the following procedures: (1) teachers introduce

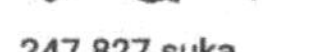

Figure 1. Sample of narrative writing in Instagram post.
(Source: https://instagram.com/humansofny?igshid=yd5wtjd9xjjp)

the Instagram to students; (2) teacher ask them to have an Instagram-account and follow the teachers Instagram account; (3) teachers introduce the narrative writing in Instagram-post; (4) teachers ask students to provide a digital photograph related to their past story to be uploaded in Instagram; (5) teachers' guide the students to analyze semiotic modes in their photograph; (6) teachers' pass question to promote the students' multimodal literacy in producing narrative writing; (7) teachers ask students to write a story related to their photo; (8) teachers' ask students to do editing (peer editing) and revising to their narrative writing; (9) teachers ask students to post their digital photograph and narrative writing and tag it to teachers' Insta-account. In presenting Instagram post both a digital photograph and narrative writing, students need to consider several things such as (1) how word and image are combined to convey meaning; (2) how the choice of design elements draw our attention; and (3) the effect created by combining different modes of communication (e.g., shock, mismatch, humor).

Multimodal literacy is about recognizing the opportunities and limitations of the various resources that make sense and how they work together to create a multimodal text that is coherent and cohesive. Students need to consider how definitions are rendered through the semiotic tools and cite textual evidence to support their multimodal text understanding in order to critically view multimodal texts. Students also need to be aware of the affordance of various semiotic tools and how they can work together effectively to make sense in order to interact effectively through multimodal representations. The questions to focus the students on the multimodal text presents a guideline for students to interpret semiotic resources in digital photography in writing narratives.

Figure 2. Sample of narrative writing in Instagram post.
(Source: https://instagram.com/humansofny?igshid=yd5wtjd9xjjp)

This activity is assumed to promote students' multimodal literacy, where the semiotic resources are well interpreted to establish a story.

5 CONCLUSION

Instagram-posts (the digital photograph and video) are the source of modes in color, visual or images, space, gestures, etc., that can be used as media to develop university students' multimodal literacy. These excellent features can provide a stimulus for developing multimodal literacy while enhancing the students' writing skills in narrative writing. The students' ability in drawing interpretation based on the offered semiotic modes presented in Insta-post for making meaning is demanded. A critical analysis is essential in interpreting the semiotic modes. Based on the Insta-post, narrative writing shows the emotional response that lies behind the images. Therefore, the existence of Instagram as a digital educational platform assists teachers to promote ELLs' multimodal literacy in narrative writing under TPACK-21CL.

REFERENCES

Abney, A. K., Cook, L. A., Fox, A. K., & Stevens, J. (2018). Intercollegiate Social Media Education Ecosystem. *Journal of Marketing Education,* 41(3), 254–269.

Binder, M. & Kotsopoulos, S. (2011). Multimodal Literacy Narratives: Weaving the Threads of Young Children's Identity Through the Arts. *Journal of Research in Childhood Education, 25:4.*
Budge, K. (2020). Visually Imagining Place: Museum Visitors, *Instagram*, and the City. *Journal of Urban Technology,* 27(2), 61–79.
Burgess, J. & Rowsell, J. (2020). Transcultural-affective flows and multimodal engagements: reimagining pedagogy and assessment with adult language learners. *Language and Education,* 34(2), 173–191.
Ellison, E. (2017). The #Australian Beachspace Project: Examining Opportunities for Research Dissemination Using Instagram. *M/C Journal, 20*(4).
Friedman, A. (2018). To "Read" and "Write" pictures in early childhood: multimodal visual literacy through Israeli Children's digital photography. *Journal of Children and Media,* 13(3), 312–328.
Haggerty, M. & Mitchell, L. (2010). Exploring curriculum implications of multimodal literacy in a New Zealand early childhood setting. *European Early Childhood Education Research Journal,* 18(3), 327–339.
Jones, P., Turney, A., Georgiou, H. & Nielsen, W. (2020). Assessing multimodal literacies in science: semiotic and practical insights from pre-service teacher education. *Language and Education,* 34(2), 153–172.
Kennedy, L.M., Oviatt, R.L. & De Costa, P.I. (2019). Refugee Youth's Identity Expressions and Multimodal Literacy Practices in a Third Space. *Journal of Research in Childhood Education,* 33(1), 56–70.
Kwon, H. (2020). Graphic Novels: Exploring Visual Culture and Multimodal Literacy in Preservice Art Teacher Education. *Art Education,* 73(2), 33–42.
Lim, F.C. (2018). Developing a systemic functional approach to teaching multimodal literacy. *Functional Linguistics,* 5(13).
Loerts, T. & Heydon, R. (2016). Multimodal literacy learning opportunities within a grade six classroom literacy curriculum: constraints and enablers. *Education,* 3(13), 490–503.
Meneses, A., Escobar, J.P., & Véliz, S. (2018). The effects of multimodal texts on science reading comprehension in Chilean fifth-graders: text scaffolding and comprehension skills. International *Journal of Science Education,* 40(18), 2226–2244.
Mills, K.A., & Doyle, K. (2019). Visual arts: a multimodal language for Indigenous education. *Language and Education,* 33(1), 1–22.
Mills, K.A., Davis-Warra, J., Sewell, M & Anderson, M. (2016). Indigenous ways with literacies: transgenerational, multimodal, placed, and collective. *Language and Education,* 30(1), 1–21.
Papadopoulou, M., Goria, S., Manoli, P., & Pagkourelia, E. (2019). Developing multimodal literacy in tertiary education. *Journal of Visual Literacy,* 37(4), 317–329.
Pérez-Gómez, F.A., & Daza, C.V. (2019). Shaping Narrative Writing Skills Through Creating Picture Books. *Gist Education and Learning Research Journal*, (19), 148–171.
Pietikäinen, S. & Pitkänen-Huhta, A. (2013). Multimodal Literacy Practices in the Indigenous Sámi Classroom: Children Navigating in a Complex Multilingual Setting. *Journal of Language, Identity & Education, 12(4)*
Refaie, I.E., & Hörschelmann, K. (2010). Young people's readings of a political cartoon and the concept of multimodal literacy. *Discourse: Studies in the Cultural Politics of Education, 31:2.*
Rhoades, M., Dallacqua, A., Kersten, S., Merry, J., & Miller, M. C. (2015). The pen(cil) is mightier than the (s) word? Telling sophisticated silent stories using Shaun Tan's wordless graphic novel, The Arrival. *Studies in Art Education, 56*(4), 307–326.
Smith, A., & Anderson, M. (2018, March 1). Social Media Use in 2018. *www.pewresearch.org*
Soviyah, S., & Etikaningsih, D. R. (2018). Instagram Use to Enhance Ability in Writing Descriptive Texts. *Indonesian EFL Journal, 4(2).*
Tan, L., Chai, C.S., Deng, F., Zheng, C.P., & Drajati, N.A., (2019). Examining pre-service teachers' knowledge of teaching multimodal literacies: validation of a TPACK survey. *Educational Media International, 56:4, 285–299.*
Tan, L., Zammit, K., D'warte, J. & Gearside, A. (2020). Assessing multimodal literacies in practice: a critical review of its implementations in educational settings. *Language and Education,* 34(2), 97–114.
TrackSAFE Foundation (2014). Useful Question Prompts when Analysing Visual and Multimodal Texts. *https://tracksafeeducation.com.au/*
Tuzel, S. (2013). Integrating Multimodal Literacy Instruction into Turkish Language Teacher Education: An Action Research Study. *The Anthropologist,* 16(3), 619–630.
Zhang, L. (2013). Mobile Phone Technology Engagement in the EFL Classroom. International Conference on Software Engineering and Computer Science. *Proceedings of the 2013 International Conference on Software Engineering and Computer Science,* 171–174.

Post Pandemic L2 Pedagogy – Adi Putra & Arifah Drajati (Eds)
© 2021 Taylor & Francis Group, London, ISBN 978-1-032-05807-8

Students' perception of the use of digital comics in Indonesian EFL reading classrooms

Atika Diyah Saputri, Sunardi & Akhmad Arif Musadad
Universitas Sebelas Maret, Indonesia

ABSTRACT: This study aims to explore the students' perception of using digital comics as digital learning media. This study used a survey research design with a questionnaire given to 80 students of eleventh grade in private senior high schools in Indonesia chosen randomly. The result showed that 79.2% of students are unfamiliar with digital comics. The finding showed that 68% of the students used PowerPoint, 24% used book media, and 8% used learning videos. Students' perceptions of digital comics show that 93% of students agreed and 7% disagreed used digital comics in the L2 reading classrooms. Pedagogical implications for the integration of digital comics in L2 reading classrooms are discussed.

Keywords: Digital comics, digital learning media, extensive reading, L2 reading

1 INTRODUCTION

Current technological developments change human life. These changes can be found in the education sector, where digitization and digital tools have transformed learning processes. Comics are an example of media that has evolved and now developed into digital and interactive forms. Traditionally only available in the form of print media, recent technology has given birth to digital-based comics. As the name suggests, digital and printed comics' main difference is that the digital one can be read using electronic devices (Petersen 2010). Some of the advantages of digital comics include being more durable, cheap, interactive, and easily accessible (McCloud 2000).

The current COVID-19 pandemic has forced schools to conduct distance learning, where students study remotely from home. The concerns for the effectiveness of online learning arise. Many teachers only focus on fulfilling the material in the 2013 curriculum, not on mastering the learning material that the students must achieve. The causes of difficulty in reading learning during the pandemic were interesting. Students read less because the teacher did not support students with digital media. Digital comics can be found easily on the internet and are yet to be recognized as educational tools. No wonder that digital comics in L2 reading classrooms are not typical and even thought to be non-academic.

Digital comics have been widely used in various fields, including education. The other researchers have modified digital comics as an alternative learning media for learning objectives, subjects, and educational levels (Lazaridis et al. 2015). Some previous studies have also investigated the potentials of digital comics in L2 classrooms, according to Issa (2018). Comics are a bridge that can develop student literacy for change in the 21st century. Other researchers (Vassilikopoulou et al. 2011) found that digital comics' educational use is designed to encourage students to acquire L2 reading skills, imagination, and experiences to create digital comics stories. Then, Tajima (2017). describes users of self-learning English comic books published in Japan related to different language books. Learners sometimes agree and fight against their views about teaching English language and native English. Therefore, Users struggle with the prevalent ideology of language. Previous studies mostly focused on exploring English literacy in language learning. More

 DOI 10.1201/9781003199267-24

studies should be done in the L2 reading classroom to uncover digital comics' potentials as a digital learning media.

Therefore, in this study, I will explore digital comics as digital media in reading classes. The focus of this research will be students' perceptions of using digital comics in the reading classroom. This research will answer the following questions: first, how is a student's experience in using digital comics in L2 reading classrooms?. Second, what types of learning media do students often use in L2 reading classrooms? Lastly, how do students perceive digital comics in L2 reading classrooms?

2 LITERATURE REVIEW

Comics can be defined as "sequential art" by combining images and text (McCloud 1993). Comic print that adapts to digital form can be called a digital comic. For example, e-books comic versions (Garrish 2011). Reading is one of the necessary skills that can help students in their process of learning English. However, not all students find reading enjoyable because they might have difficulty understanding the text, main ideas, and answering questions. In this case, extensive reading of written texts helps improve students' language acceptance process (Huang 2015). An alternative term in defining extensive reading is "reading pleasure" (Mikulecky 1990). In addition to finding pleasure in reading, readers will acquire a lifelong reading habit (Jacobs et al. 1997). Digital media is used in the L2 reading classroom. Digital comics could be an attractive alternative tool application in the reading classroom.

Several researchers reported that digital comics could be potentially used in L2 reading classrooms. The teacher's comics aim to explain aspects of differences in the reading process, such as paying attention to gender in media texts, introducing dialogue, and solving problems in a text. Students' perceptions indicate that comics have unique characteristics, fun, and helpful in reading learning strategies (Hammond & Danaher 2012). Collaborative use of comics can be used in the standard curriculum and make learning better (Dallacqua 2020). Comics develop students' multimodal literacy skills and prepare for a change in the 21st century (Issa 2018). Therefore, comics can develop students' motivation and literacy skills.

The use of digital comics is an appropriate learning material in L2 reading classrooms. Therefore, observers have underlined a systematic study that links digital comics with the learning objectives to be achieved. In this study, I would like to examine digital comics in Indonesian EFL reading classrooms. Comics describe it as a creative process in finding solutions to practical problems. The educational digital comic design aims to motivate students in acquiring language skills, imagination, and cultural experiences in making stories such as multimodal comics (Vassilikopoulou et al. 2011).

3 METHODOLOGY

This research was aimed to investigate students' perceptions of adopting digital comics in the reading classroom. The study was conducted in one of the private senior high schools in Indonesia. The research participants are students in eleventh grade with a total sample of 80 students. On average, the participants are 17–18 years old, and 78% are female.

This study uses descriptive research. Descriptive research is explaining the characteristics of a particular event. It is more concerned with what or why something happened (Nassaji 2015) with the survey as the data collection method. The data collection uses a questionnaire that is distributed to student population samples. The collecting data technique is by distributing questionnaires and then analyzing the results quantitatively using percentages and analysis tools. According to Brown (2001, p. 6), a questionnaire is a written instrument in the form of a question or statement that the respondent must answer by writing down or choosing an existing answer. The study began with a sample questionnaire for 80 students. This research only focuses on students' perceptions of digital comic learning media in the reading class. It makes the composition of the questions in the questionnaire consisting of the types of digital learning media commonly used by students,

students' interest in using digital learning media, and perceptions about one of the digital learning media in the form of a digital comic application, and the last question about the reasons for students choosing digital comics as learning. After the sample questionnaire was distributed, the data are analyzed quantitatively using percentages.

4 RESULTS AND DISCUSSION

This part will show the research results related to students' experience on the use of digital comics, types of learning media, and students' perceptions of digital comics. This section will describe the usage of digital comic media used during the L2 reading classroom.

4.1 *Students' experience of the use of digital comics*

In this section, the researcher will explain students' experiences related to digital comic media. As shown in Figure 1, the students' findings filled out by the students indicated that 79.2% of students were unfamiliar, and 20.8% were familiar with digital comics media. Students admitted that their teacher had not applied digital comics media in the L2 reading classroom. It means the students did not have a learning experience using digital comics as learning the media during these online learning classes.

Teachers should be able to adopt digital media in online teaching during the COVID-19 pandemic. This condition requires science and skills, and trust in online teaching success (Konig et al. 2020).

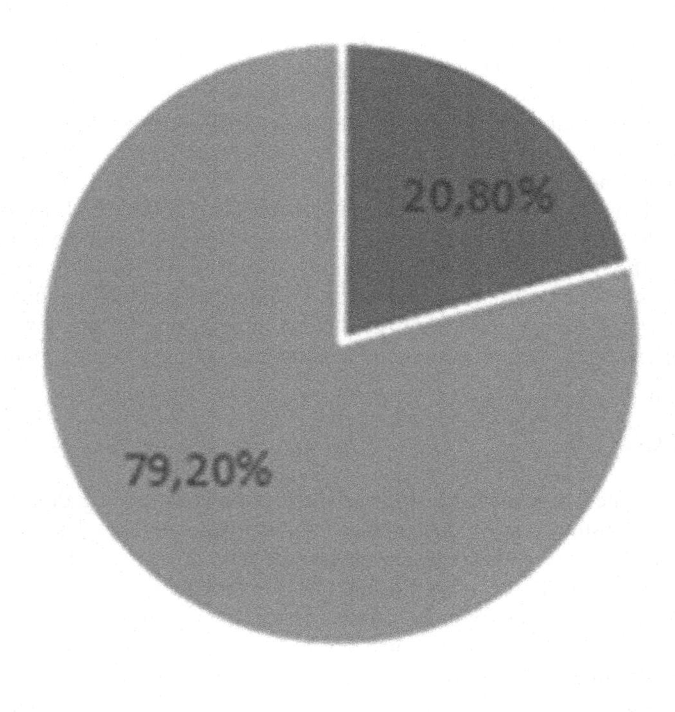

Figure 1. The students' familiarity with the use of digital comics.

Teaching adaptiveness is considered a determinant of the quality of education. This finding has followed research that emphasizes the meaning and competence of a teacher's quality in achieving educational goals (Kaiser & Konig 2019). Based on the previous study, digital comics' educational use is designed to encourage students to acquire language skills (Vassilikopoulou et al. 2011) and an accessible gateway to text-based literacies (Issa 2018).

4.2 *Types of students' learning media*

In this section, the researcher will explain the digital learning media that students use in reading classrooms. As shown in Figure 2, the questionnaire results showed the media that is often used is PowerPoint at 68%, book 24%, and video 8%. Teachers use PowerPoint to deliver learning material, while students believe this media is easy to make and does not require a lot of time to take.

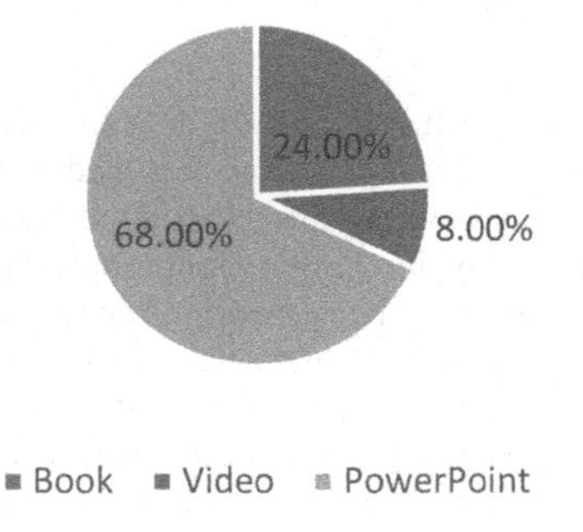

Figure 2. Types of learning media.

The students' interview was found that most of the teachers used other digital learning media. It can be affected in the L2 reading classroom because the students have difficulties in accepting the materials presented by the teacher. Sometimes, the students feel bored because the learning media used is not suitable for the topics. The most potential media during online learning is digital learning media. The evidence showed digital technology enables learning for meta-analysis (Chauhan 2017). According to previous research, technology encourages teachers to facilitate learning, and teacher actions can improve student conceptual understanding and learning choices (Engeness 2020).

4.3 *Students' perception of the use of digital comic*

As shown in Figure 3, Students' perceptions of digital comics media showed that 93% of students agree to adopt digital comics in the reading classroom. In comparison, only 7% of students disagree. According to their responses, the students had never used this type of learning media in the L2 reading classrooms.

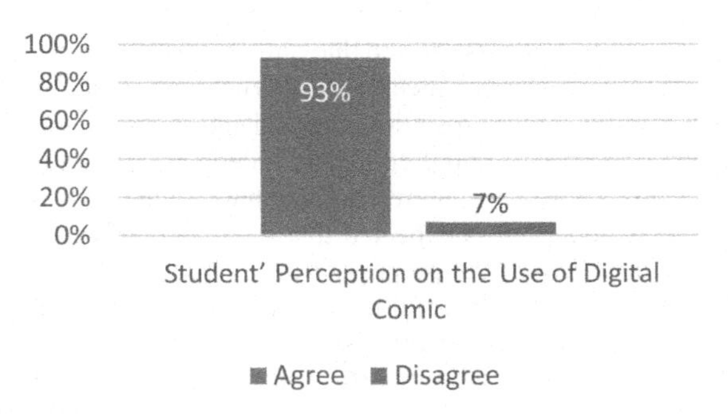

Figure 3. Student' perception of the use of digital comic.

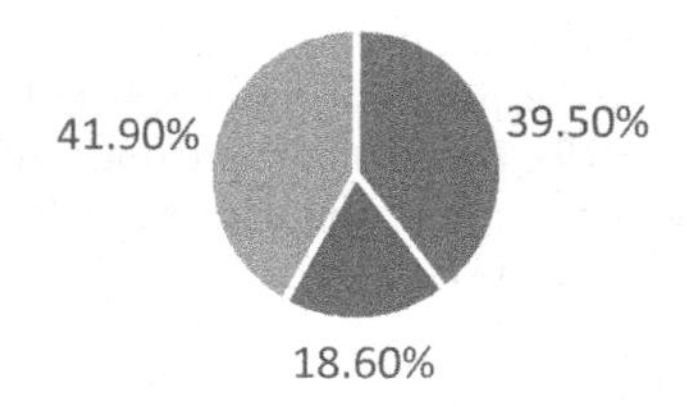

Figure 4. Students' reason for choosing digital comic media.

As shown in Figure 4, the data findings related to digital comics were supported by several reasons for students choosing digital comics as digital learning media during the COVID-19 pandemic. Students admit that digital comics have an attractive appearance. It indicated that comics are different from other learning media, which have a monotonous appearance. 18.6% of the students admitted that digital comics are interactive because digital comics contain stories expressed through animation and pictures. Digital comics are a learning media that produces interaction, and 41.9% of students admit digital comics have various materials and information. The information and material displayed in digital comics are adjusted to be discussed. The selection of characters and background of comic animation settings also affects the process of learning. The use of digital media is opening new possibilities in learning and also having to learn something about media and how to operate it. Benefits of digital media include making it easier to complete tasks, being flexible, and carrying out learning visually (Henderson et al. 2017). According to previous research, Collaborative use of comics can be applied to the standard curriculum and can better change student learning structures (Dallacqua 2020).

The students unfamiliar with digital media as learning media indicate no students' experience using digital comics. Most of the teachers used other media. However, the students have difficulties in the L2 learning classroom. It concluded that digital comics as learning media are potentially used in L2 reading classrooms. The data finding was supported by several reasons for students choosing digital comics as digital learning media during the COVID-19 pandemic. Comics have an attractive appearance, interactive, and a variety of materials and information.

5 CONCLUSION

In general, this study tried to describe the students' perception of digital comics in reading. Based on this study, 79.2% of students are unfamiliar with digital comics as reading classroom media. Learning media that was used by students showed that online learning currently, 8% of students use book media, 24% of students use learning videos, and 68% using PowerPoint. Students' perceptions of digital comic media show that 92.9% of students agreed and 7.1% disagreed used digital comics in the reading classroom. The use of digital comics is indeed a high potential for improving the classroom's quality and effectiveness. Some researchers have found that using digital comics for learning will be useful for students. It is expected that the results of this research can become a consideration for the sustainability of the process of developing learning media products, given that students are still unaware of learning multimedia. Meanwhile, several relevant studies have proven that digital comics in the learning process positively impact student performance. The pedagogical implication that the available support resources can facilitate teachers in creating an optimal learning process. There needs to be an awareness of teachers about the selection of resources used in the learning process. Teachers must be able to collaborate in digital technology to shape students' conceptual understanding. However, our findings are based on narrow observations, so this area deserves further research.

REFERENCES

Brown, H. G. (2001). Teaching by principles: Interactive approach to language pedagogy. New York, *NY: Pearson Longman.*

Chauhan, S. (2017). A Meta-analysis of the impact of technology on learning effectiveness of elementary students. *Computers & Education*, 105, 14–30.

Dallacqua, A. K. (2020). Reading comics collaboratively and challenging literacy norms. Literacy *research and instruction, 59*(2), 169–190.

Engeness, I. (2020). Developing teachers' digital identity: towards the pedagogic design principles of digital environments to enhance students' learning in the 21st century. *European Journal of Teacher Education,* 2(1), 1–13.

Garrish, M. (2011). What Is EPUB 3?. 1st ed. Beijing: *O'Reilly Media.*

Hammond, K., & Danaher, K. (2012). The value of targeted comic book readers. *ELT journal*, 66(2), 193–204.

Henderson, M., Selwyn, N., & Aston, R. (2017). What works and why? Student perceptions of 'useful digital technology in university teaching and learning. *Studies in Higher Education, 42*(8), 1567–1579.

Huang, Y. C (2015). Why don't they do it? A study on the implementation of extensive reading in Taiwan. *Cogent Education*, 2(1), 1–13.

Issa, S. (2018). Comics in the English classroom: a guide to teaching comics across English studies. *Journal of Graphic Novels & Comics, 9*(4), 310–328.

Jacobs, G. M., Davis, C., & Renandya, W. A. (1997). *Successful strategies for extensive reading* (1st ed.). Singapore: SEAMEO Regional Language Centre.

Konig, J., Jäger-Biela D. J., & Glutsch, N. (2020). Adapting to online teaching during COVID-19 school closure: teacher education and teacher competence effects among early career teachers in Germany, *European Journal of Teacher Education.* 43(4), 608–622.

Lamminpaa, J., Vesterinen, V. M., & Puutio, K. (2020). Draw-a-science-comic: exploring children's conceptions by drawing a comic about science. *Research in Science & Technological Education,* 1–22.

Lazaridis, F., Mazaraki, A., Verykios, V., & Panagiotakopoulos, C. (2015). E-comics in teaching: Evaluating and using comic strip creator tools for educational purposes. *2015 10th International Conference on Computer Science & Education*, 305–309.

McCloud, S. (1993). *Understanding Comics*: The Invisible Art by McCloud, Scott. New York, NY: Harper Collins.

McCloud, S. (2000). *Reinventing Comics*. 1st Perennial Ed ed. New York, NY: Harper Collins.

Mikulecky, B. S. (1990). A *short course in teaching reading skills* (1st ed.). Reading, MA: Addison-Wesley.

Nassaji, H. (2015). Qualitative and descriptive research: Data type versus data analysis. *Editorial Language Teaching Research*, 19(2), 129–132.

Petersen, R. (2010). Comics, manga, and graphic novels. History of Graphic Narrative. Santa Barbara, California: *ABC-CLIO.*

Tajima, M. (2017). Weird English from an American'? Folk engagements with language ideologies surrounding a self-help English language learning comic book published in Japan. *Asian Englishes. 20(1)*, 65–80.

Vassilikopoulou, M., Retalis, S., Nezi, M., & Boloudakis, M. (2011). Pilot use of digital educational comics in language teaching. *Educational Media International*, 48(2), 115–126.

Post Pandemic L2 Pedagogy – Adi Putra & Arifah Drajati (Eds)
© 2021 Taylor & Francis Group, London, ISBN 978-1-032-05807-8

Does Google Docs facilitate collaborative writing? A case from Indonesia

Anis Handayani, Nur Arifah Drajati, Ema Wilianti Dewi & Niken Sri Noviandari
Universitas Sebelas Maret, Indonesia

ABSTRACT: This qualitative case study aims to explore metadiscursive Google Docs in facilitating writing practice in one Indonesian university. This study was conducted among 37 third-semester students in an academic writing class. To explore this issue, we qualitatively analyzed the teacher's and students' comments and text revision in Google Docs, as well as their follow-up interviews. The results show that Google Docs could engage students and teachers to give and respond to feedback. Furthermore, despite its challenges, the students showed positive views of Google Docs as a collaborative writing aid. It implies that Google Docs did support collaborative writing, but the students and teachers need more exposure to use it effectively. Educational practitioners may provide such training to support writing practice, specifically collaborative writing, in this online learning era.

Keywords: academic writing, collaborative writing, computer-assisted feedback, online learning

1 INTRODUCTION

The writing process has become a trend in recent years changed the paradigm of conventional approaches that emphasize students' final writing. Vygotsky's (1978) sociocultural theory was applied to EFL writing classrooms that focus on the teachers' supports and feedback (Aljaafreh & Lantolf 1994) and peer supports and feedback in a successful contribution (Villamil & Guerrero 2016). As many researchers found, feedback and collaborative writing have a positive impact on students' writing improvement (Chen & Yu 2019; Yang 2017).

This current study tries to connect the sociocultural theory with writing, specifically online collaborative writing. Storch (2011) argued that collaborative writing (CW) refers to the co-construction or co-authorship of a text. By sociocultural theory, students can achieve ZPD through interaction in collaborative writing. It combines the benefits of learner interaction with the recursivity that writing encourages, such as testing hypotheses, receiving and noticing feedback, and focusing on accuracy (Abrams 2019). In collaborative writing, interaction becomes the central point students need to be engaged (Wang 2019). This Google doc engages in giving and responding to feedback (Alharbi 2019).

However, each student has different characteristics. Some may have confidence in giving feedback depending on many factors such as learning culture, linguistic proficiency, etc. As Kale (2014) revealed, some students are reluctant to edit or comment on their peer's writing during online collaborative writing, and it affects the interaction in an online collaborative writing activity. However, Alharbi (2019) found that the students in Saudi Arabia were quite interested in giving feedback to their peers' writings and frequently responding to the instructor's feedback. Thus, this current study explores how Google Docs facilitates collaborative writings and how Indonesian students and lecture views this teaching aid as Indonesian and Saudi students indeed have different characteristics.

DOI 10.1201/9781003199267-25

Table 1. The students' group.

Pairs	Student Pseudonyms
G1	S1-G1 & S2-G1
G2	S1-G2 & S2-G2
G3	S1-G3 & S2-G3
G4	S1-G4 & S2-G4
G5	S1-G5 & S2-G5
G6	S1-G6 & S2-G6
G7	S1-G7 & S2-G7
G8	S1-G8 & S2-G8
G9	S1-G9 & S2-G9
G10	S1-G10 & S2-G10
G11	S1-G11 & S2-G11
G12	S1-G12 & S2-G12
G13	S1-G13 & S2-G13
G14	S1-G14 & S2-G14
G15	S1-G15 & S2-G15
G16	S1-G16 & S2-G16
G17	S1-G17 & S2-G17
G18	S1-G18, S2-G18, & S3-G18

2 RESEARCH METHOD

This qualitative case study is situated in one Indonesian university that embraces 30,000 students with numerous undergraduate, graduate, and doctorate study programs. The 37 second-year students joined the English Department as the participants for this study. The students were between 19 and 20 years old, 21 females and 16 males, with an intermediate writing proficiency level. The participants were divided into 18 groups, each group consisting of 2–3 students. All students are pseudonyms to keep anonymity and privacy. Table 1 shows the students' groups.

The participants were given a task to write an article review from reputable journals. They did this task in three months through Google Docs and with the lecturer's supervision. They wrote and edited their writing based on the peers' and lecturer's given feedback. Their writings and their Google docs activities were then observed and analyzed to find out how it engaged both students and lecturers to give writing feedback and respond to such feedback.

When the students finished their writings, they were given a written interview through Google form to find out their writing experience collaboratively using Google Docs. Six selected students were then in-depth interviewed through phone calls to give additional information. The data, both written interviews, and phone recordings were then analyzed under two themes, affordance and weakness of using Google Docs. Then, the data were synchronized between two researchers to conclude.

3 FINDINGS AND DISCUSSION

3.1 *Metadiscursive Google Docs in facilitating collaborative writing*

3.1.1 *Engaging teachers and students in feedback*

Our study results show that both the teacher and students make good use of feedback in this collaborative writing. Their feedbacks were targeted on two issues, global and local (see Table 2). Table 3 shows the percentage of the teacher and students' feedback based on the target issues.

The use of Google docs was proved to engage the teacher and students in giving feedback. Table 3 shows feedback on global issues (47, 51%) outnumbered feedback on local issues (45, 49%).

Table 2. Feedback samples focusing on global and local issues.

	Teacher feedback	Student feedback
Global issue	**Lecturer**: Please write a research method	**S2-A9**: There are only four sentences, better to add some more
Local issue	**Lecturer**: Check singular or plural	**S2-B1**: Delete s

Table 3. Percentage of teacher and student feedback.

	Teacher feedback	Student feedback	Total
Global issue	15 (16%)	32 (35%)	47 (51%)
Local issue	42 (46%)	3 (3%)	45 (49%)
Total	57 (62%)	35 (38%)	92 (100%)

This finding is in line with Alharbi's (2019). It indicates that most issues detected are global issues related to the content and organizational issues. It also shows that local issues such as grammar and spelling were used correctly, so there was less feedback. Also, it reveals that Indonesian students tend to ignore global issues such as the writing organization that there are more errors on this matter than another.

Comparing the teacher's and students' feedback, the teacher gave more feedback than the students, which is in contrast to Alharbi's (2019). It may indicate that the students here tend to be afraid to give feedback. They tend to wait for the teacher's feedback rather than give feedback. They were afraid that their feedbacks were incorrect and making their peers' writings worse. They thought they were not competent enough to give feedback to other students who had the same competence level. It is relevant to EFL students' culture in which they feel inferior over their capability.

Furthermore, it is interesting to find out that the teacher focused their feedback on local issues (42, 46%) more frequently than on global issues (15, 16%). In contrast, the students focus on global issues (32, 35%) more frequently than local issues (3, 3%). It indicates that students view the writings broadly as a whole text to give feedback on global issues more frequently (see Table 2). They did not have confidence in their grammar or sentence structure competence, so they tend to avoid giving feedback on this matter.

3.1.2 *Engaging students in responding to feedback*

Unlike in offline written feedback, the students here can respond to the lecturer's feedback directly and immediately. This Google doc provides a medium for the students and the lecturer to give feedback and respond to it directly. Figure 1 shows a screenshot of a student's response to the lecturer's feedback.

The use of Google Docs here indeed makes differences from offline feedback in which the student only revise their writings without responding to the feedback. This finding supports Ebadi and Rahimi's (2017) and Strobl's (2014). However, this activity of responding to such feedback was rarely found here. Many students only corrected their writings without responding to the feedback as in offline feedback. They were hesitant to ask more about the feedback since they probably understood the feedback well. Some students said that they were afraid to respond to such feedback as they usually did not do it in offline feedback. It has been a habit of the students that they only revise without responding to the feedback.

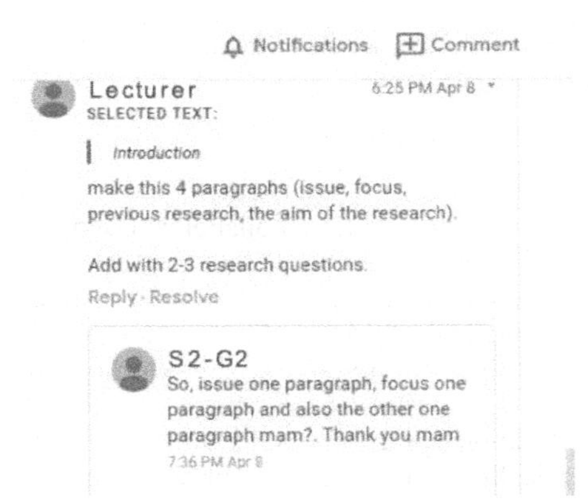

Figure 1. A screenshot of a student's response to the lecturer's feedback.

3.2 *Students' views of Google Docs for collaborative writing*

Mostly the students here agreed that Google Docs facilitate their collaborative writings. They confessed that Google Docs supported peer editing, making it easier for them to create, view, edit, and check their writings anytime. This finding agrees with Ebadi and Rahimi's (2017) and Alharbi's (2019). The students confessed that they could supervise their peers' works directly.

> "Effective, because it is done online and each other can create, view, edit or check the results of activities in one document." (S2-G1)

Another affordance of using Google Docs is that their writing editing is traceable, as found by Ebadi and Rahimi (2017). They can quickly locate the part they revised to undo it or change it. It was an essential feature of Google Docs, which they did not find in offline writings. This feature indeed has a positive impact on students' peer editing and feedback.

> "We can write online, and all the changes we made can be traced. We can also review changes via history." (S1-G7)

Besides the affordances, this teaching aid indeed has weaknesses. The first one is the technological barrier. It commonly happens to those who live in the countryside, as Ramos et al. (2011) found, related to the poor internet connection. The students said they liked using Google docs to write collaboratively, but the unstable connection frequently impeded them. It indicates that this teaching aid probably provides a better learning experience and outcomes to those who live in areas with a good internet connection.

> "Yes, unstable connection. Because …I live in a small village." (S2-G5)

Another challenge is the students' lack of knowledge and experience in using Google Docs. Many students confessed that it was their first time using Google Docs. This finding turns out to be different from Ebadi and Rahimi's (2017) as the students there were already trained on how to use Google Docs. The students here probably never use this kind of teaching aids in their prior learning experience. Some did not even know if this Google doc has a chat feature. They frequently have the discussion using different Apps when they actually can do it directly in Google docs.

> "I do not really understand the features." (S2-G2)

Despite the challenges of using Google docs for collaborative writing, the students mostly prefer using Google (24 out of 37 students). They confessed that it was more effective than offline feedback since it did not require face-to-face and take much time. Also, they can edit the writings directly, and the edit they made is traceable. It is undoubtedly one of the advantages of using Google Docs, which cannot be provided by offline feedback.

"For collaborative writing, I prefer Google docs, because it's more effective." (S1-G10)

4 CONCLUSION

This study reveals that Google Docs facilitates collaborative writing for Indonesian students. It supports students' and lecturer's engagement in giving and responding to feedback. However, in responding to the lecturer's feedback, the students were reluctant since they usually only corrected their writings without responding to the feedback. Furthermore, despite its challenges, the students here have positive views of Google docs for collaborative writings. They liked using Google Docs for collaborative writing since it was effective.

This study implies that educational practitioners may use this teaching aid to support the teaching and learning process, specifically in collaborative writing. However, it should still be fitted to the students' conditions to avoid the challenges. Furthermore, the teacher's views on this matter were not explored yet. Thus, future studies may focus on teacher's views to explore this matter further. A similar study with different students and different conditions may be worthwhile to conduct to find out the best condition to use Google Docs as a teaching aid for collaborative writing.

REFERENCES

Abrams, Z. I. (2019). Collaborative writing and text quality in Google Docs. *Language Learning & Technology, 23*(2), 22–42.

Alharbi, M. A. (2019). Exploring the potential of Google Doc in facilitating innovative teaching and learning practices in an EFL writing course Exploring the potential of Google Doc in facilitating innovative teaching and learning practices in an EFL writing course. *Innovation in Language Learning and Teaching, 0*(0), 1–16.

Aljaafreh, A. L. I., & Lantolf, J. P. (1994). Negative feedback as regulation and second language learning in the Zone of Proximal Development. *The Modern Language Journal, 78*(4).

Chen, W., & Yu, S. (2019). Implementing collaborative writing in teacher-centered classroom contexts: student beliefs and perceptions. *Language Awareness, 28*(4), 247–267.

Ebadi, S., & Rahimi, M. (2017). Exploring the impact of online peer-editing using Google Docs on EFL learners' academic writing skills: a mixed-methods study. *Computer Assisted Language Learning, 30*(8), 787–815.

Kale, U. (2014). Can they plan to teach with Web 2.0? Future teachers' potential use of the emerging web. *Technology, Pedagogy, and Education, 23*(4), 471–489.

Ramos, F., Tajú, G., & Canuto, L. (2011). Promoting distance education in higher education in Cape Verde and Mozambique. *Distance Education, 32*(2), 159–175.

Storch, N. (2011). Collaborative Writing in L2 Contexts?: Processes, outcomes, and future directions. *Annual Review of Applied Linguistics, 31*, 275–288.

Strobl, C. (2014). Affordances of web 2.0 technologies for advanced collaborative writing in a foreign language. *CALICO Journal, 31*(1), 1–18.

Villamil, O. S., & Guerrero, C. M. De. (2016). theory: A framework 2 sociocultural for understanding the social-cognitive dimensions of peer feedback. In *feedback in second language writing* (pp. 23–41). Cambridge University Press.

Vygotsky, L. (1978). *Mind in Society: The development of higher psychological processes*. Harvard University Press.

Wang, L. (2019). Effects of regulation on interaction pattern in a web-based collaborative writing activity. *Computer Assisted Language Learning, 0*(0), 1–35.

Yang, Y. F. (2017). New language knowledge construction through indirect feedback in web-based collaborative writing. *Computer Assisted Language Learning, 31*(4), 459–480.

Post Pandemic L2 Pedagogy – Adi Putra & Arifah Drajati (Eds)

Test-repeaters' perceptions of difficulty on the TOEFL listening test

Anang Widodo, Novia Russilawatie & Septi Riana Dewi
Universitas Teknologi Yogyakarta, Indonesia

ABSTRACT: This case study investigates the test-repeaters' perceptions on the TOEFL Listening test. The participants were students in a private university in Indonesia. Test-takers' scores were analyzed and coded based on the name and the lowest score in each section. We collected the data of test-repeaters' perceptions from a semi-structured interview. The interview transcript was analyzed, coded, and triangulated with test-repeaters' scores. The results indicate that the participants believed that the TOEFL test was important, but it was challenging. Most participants failed on the listening sections due to a lack of preparation and problems relating to vocabulary, including pronunciation and accent used in the audio. This study emphasizes that having good planning and preparation is essential to avoid failure. This result carries crucial pedagogical implications for test-takers, TOEFL trainers, and policymakers.

Keywords: Difficulty in TOEFL test, test-repeaters, TOEFL listening test

1 INTRODUCTION

The use of the TOEFL test as a standardized English test has been applied at several universities (O'Dwyer et al. 2018; Ginther & Elder. 2014) Implementing TOEFL as the standardized test at university brings positive and negative impacts for students as presented by the following studies. Hung and Huang (2019) mentioned that students believed that campus-wide standardized tests benefit their English proficiency when they work in the future. Besides its benefits, this implementation also results in problems for students who cannot meet the minimum standard score. They, consequently, need to retake the same test. The study of test-takers who should retake the test was conducted by Cho and Blood (2020) and Knoch et al. (2020). These studies investigate the preparation that test-takers made before the test and their score changes. It indicates that repeaters should spend more time to focus on test preparation before the test day to achieve their targeted scores.

One of Indonesia's private universities implements an English standardized test for its students as one of the graduation requirements. Students should take the test at the university's education, certification, and training center. The minimum score that students should get is 450–500, depending on the minimum standard score set by each faculty. For many test-takers, the minimum standard score set is considered high. Many of them failed to meet the minimum standard score set by their college. They, therefore, had to retake the test several times until they achieved the score required. The number of repeaters is recently increasing since some faculties raise the minimum standard score for their students. During the period of the study, there were 1,690 test-takers registered and 592 test-takers or 35% of the test-takers failed to meet the minimum standard score and should retake the test several times.

A standardized English language test as a requirement for graduation is also implemented in more than 90% of universities in Taiwan (Wu and Lee 2017). This study showed that the implemented policy gets many criticisms because it is considered that the test cannot improve students' English language proficiency. However, students have positive perceptions of this English graduation benchmark policy. The study conducted by Li et al. (2012) showed that the CET (College

DOI 10.1201/9781003199267-26

English Test), an English standardized test designed to measure Chinese undergraduate students' English proficiency, motivated test-takers to make a more significant effort to prepare for the test. They eagerly put more effort to learn English skills tested in CET. These two previous studies are relevant to the current study because they have similar test-takers' perspectives toward the university's standardized test. The regulation implemented impacts students as test-takers because they have to take this test. The previous studies proved that this policy positively influences students to study harder to improve their English skills to meet the standard score. However, those who should retake the test several times possibly will find it stressful and exhausting. Thus, this current study investigates perceptions of test-takers who have failed several times to meet the minimum standard test score set by faculty as a requirement for their graduation.

Although there are some researches about test-takers' perceptions (Galikyan et al. 2019; Fan 2014; Pearson 2019, Hamid et al. 2019; Wu and Lee 2017; Li et al. 2012), there is no research focusing on test-takers' perception toward difficulty on the TOEFL Listening test. This current study focuses on investigating this issue. It is essential to investigate this issue to raise test-takers' awareness about the large number of test-takers who should retake the test and their problems on the test. Besides, it hopefully gives policymakers new insights to make some regulations that address the TOEFL test repeaters. This study investigates the most challenging section in the TOEFL test and test-takers' perceptions of their TOEFL test failure.

2 LITERATURE REVIEW

Listening is one of the skills tested in a standardized test. Chang et al. (2013) mention that students face some problems such as fast speech rates, accent, connected speech, colloquial usages and slang, and transient information. Brindley and Slatyer (2002) concluded some factors affecting the difficulty of listening comprehension tasks into three categories. They are as follows, 1) the nature of the input, including vocabulary, speech rate, accent, amount of redundancy, length of the passage, syntactic complexity, propositional density, discourse structure, noise level, register; 2) the nature of the assessment task. This category includes clarity of instructions, amount of context provided, availability of question preview, and response format, and 3) the individual listener factors. This category comprises memory, knowledge, interest, motivation, and background. In terms of the Listening section in the TOEFL test, Kostin (2004) investigated the difficulty Listening section, particularly in the difficulty of TOEFL dialogue items. The result showed that test-takers find it difficult at 1) when the dialog consists of two or more negatives, 2) when the meaning is not stated indirectly and test-takers should draw an inference from the dialogue, and 3) utterances patterns in the dialogue.

The following studies focus on test-takers perceptions of the standardized test. Galikyan et al. (2019) figured out that student test-takers have positive perceptions of the test because it can show their English language abilities. Fan (2014) mentioned that VET (Versant English Test) receives positive perceptions from its test-takers since it is reflected in test-takers' proficiency in spoken English. Some test-takers' negative perceptions toward the standardized test can be seen from the following studies. Pearson (2019) highlighted that test-takers have negative perceptions of the speaking and writing assessment accuracy in IELTS. They mistrust single examiner marking. They request detailed feedback on their test performance and an explanation about their incorrect answers. In line with the previous study, Hamid et al. (2019) also found negative perceptions of test-takers towards the standardized test that they took. Although they believed that the IELTS test was fair, they assumed that it did not provide an accurate measure of test-takers' proficiency and did not test test-takers' actual language proficiency. The study conducted by Tsai and Tsou (2009) revealed that instead of focusing on enhancing students' English competency, implementing standardized ELP tests as a graduation benchmark requirement turns the teaching and learning process into test-oriented. Besides, most participants mentioned that the standardized test implemented did not assess their learning outcomes.

Test-takers' perceptions toward the test are influenced by many factors such as English proficiency, cultures, socio-economic backgrounds, and English learning conditions Pham and Bui (2019). This study showed that the Vietnamese test-takers who live in the South were more supportive of implementing the ELPTs (English Language Proficiency Tests) than test-takers in Central Vietnam because they believed that the test motivated them to learn and improve their English proficiency. They also agree that the test can enhance competitiveness in their future workplace and further education. Such perspective reflects that students from developing and urban contexts (South) acknowledge the ELPTs (English Language Proficiency Tests), the policy, and English language skills are right for their future. Although several studies about test-takers' perceptions have been conducted, there is no study about test-takers' perceptions in the Indonesian context. This study, thus, investigates test-takers' perceptions towards their failure in the TOEFL test as a requirement for graduation in one of the private universities in Indonesia.

3 RESEARCH METHOD

The data of test-takers' scores were collected from all test-takers who registered at the university's education, certification, and training center in a one-year duration. To get the data, the researchers sent a consent letter to the head of this department. After getting permission, the researchers received the data of test-takers scores and used it as one of the main data for this study. All data used in this study will remain confidential. During the study, there were 1.690 test-takers registered. However, only 1.098 test-takers or 65% of the total participants could meet the minimum standard score from their faculty, while 592 test-takers or 35% of test-takers had to retake the test twice to seven times. Of 592 repeaters, 16 test-takers should retake the test more than 5 times. For data of the interview, the researchers selected participants through the purposive random sampling from those who failed the test more than 5 times. After selecting the participants, the researchers contacted them to ask their permission and willingness to have an interview. In the beginning, the researchers contacted 9 participants, but 3 of them did not give any responses. The participants involved in a semi-structured interview were 6 test-takers aged 22–23 years old consisting of 1 female and 5 male students. After getting permission from the participants, an interview schedule was set and it was conducted online by calling each participant through WhatsApp.

L2 teaching and learning complexity require an in-depth study to understand students and their behaviors and experiences (Nassaji 2015). A Case Study design is more appropriate to figure out this complexity since it deals with a specific case on specific participants. Yin (2017, p.14) stated that "A Case Study is an empirical method that investigates a contemporary phenomenon (the "case") in depth and within its real-world context, especially when the boundaries between phenomenon and context may not be clearly evident." This study explores test-takers' perceptions of their experience after failing the TOEFL test several times and finding out why the Listening section fails them to achieve the targeted score. The data used are the test-takers' scores in a one-year duration and interview. Test-takers' scores were analyzed and coded based on the lowest score in the Listening section. To get the data of test-takers' perceptions, a semi-structured interview was conducted. The interview transcript was analyzed and given some codes and it was triangulated with the data of TOEFL test-takers' scores. A further follow-up interview was conducted with participants to confirm and clarify the data.

4 FINDINGS AND DISCUSSION

All participants have positive perceptions of the test. They believed that TOEFL is essential for their future. This finding is in line with the study conducted by Pham and Bui (2019), which showed that implementing the English standardized test motivated test-takers to improve their English proficiency. In this study, test-takers believed that it could enhance competitiveness in their future workplace and further education. However, the current study results show that participants admitted that it is tough to get the score targeted in the TOEFL test. They found it challenging, considering that they needed 6 to 7 times to get the minimum standard score. The TOEFL test is regarded as a tough test in Indonesia and Japan (Koizumi & Nakamura 2016). This study showed many test-takers in Japanese universities need to retake the test more than once. The participants considered that the standardized test implemented was too tricky.

Based on the interview, four out of six participants agreed that listening was the most challenging section. This result is in line with a study conducted by Chang et al. (2013). This study involving low-level learners revealed that 73% of the participants perceived English language listening as difficult. In terms of difficulty faced during the test, several participants mentioned that they were not familiar with speakers' pronunciation, accent, and vocabularies used in the audio during the listening section. The data show that lack of vocabulary mastery affects test-takers' listening performance. DM realized that listening was his most difficult section since he kept failing in this section. He mentioned, "*In the Listening section, mostly I could not understand the conversation. I do not know the vocabulary used. The accent they use is very different from Indonesian people"*. As stated in the previous studies, vocabulary is one of the problems faced by students in listening (Chang et al. 2013; Brindley and Slatyer 2002) and in the TOEFL test (Kostin 2004)

The participants' other problems are the quality of the audio and listening tool used in the test and the speed of audio used. DW and DA lacked focus because the audio was not clear. It dealt with the headphones they used. DW stated, "*The speakers' voice in the Listening section is not quite clear due to the quality of the headphones used."* Fast speech rates, accent, vocabulary, length of the passage, and noise level are common problems students face in listening (Chang et al. 2013; Brindley & Slatyer 2002). NA mentioned that "*speakers speak too fast. I do not understand what they are talking about"*. NV argued, *"I do not understand TOEFL. I do not understand the listening."* Faster rate effects reduce students' comprehension in listening (Brindley &Slatyer 2002). Kostin (2004) mentioned that the dialogue's utterance patterns are some difficultietheby test-takers face in the listening section.

Another challenging aspect of listening is the ability to remember information from the audio. DM stated, "*It is difficult to understand and remember all information in the Listening section, so I answer it based on my feelings and clues from a piece of information I get.*" Namaziandost et al. (2018) stated EFL listening comprehension can be influenced by working memory since it depends on the storage and processing of information in the mind. Brindley and Slatyer (2002) categorized this difficulty into individual listener factors in which test-takers memory influences their performance in the listening test.

The last issue raised by participants in this study is the lack of preparation before the test. During the period of the study, the institution did not provide test-preparation classes for test-takers. Therefore, all participants prepared it by themselves. No one of them went to an English course/training center or hired English tutors to help them prepare for the test. Chou (2015) argued that test-takers who are categorized as 'average' and 'weak' listeners tend to ignore the importance of training in other English skills, particularly in listening. Test preparation positively influences test-takers' score gain (Liu 2014; Mickan & Motteram 2008) Pearson (2019) mentioned that 40% of candidates fail the IELTS test due to lack of preparation. Pan (2016) argued that the amount of time spent on language learning before a test is significant in determining student – score improvement. Thus, it can be offered that a lack of preparation can be one of the causes of repeaters' failure in the test.

5 CONCLUSION

All participants in the study agree that TOEFL is essential for them, but it is challenging. Most participants failed in the listening section. They all mentioned that vocabulary mastery, pronunciation, working memory, dysfunctional headset, and lack of preparation are the problems they faced. Although they are aware that their English proficiency level is not good enough, they did not do much effort to prepare for the test well. Having good planning and preparation before taking the TOEFL test is essential to avoid taking the test several times to achieve the targeted score. Consequently, they should retake the test 6 to 7 times to meet the minimum standard score set by their college. This study's results contribute to pedagogy, particularly for teachers and TOEFL instructors to explore new directions to teach TOEFL and focus on the listening section. It is essential to improve test-takers vocabulary mastery during TOEFL training. For test-takers, good preparation before the TOEFL test will give a greater possibility to achieve the targeted score. This study also suggests that the university provides test-preparation classes for students or, particularly repeaters, to improve their test scores and avoid retaking the test several times. This study deals with the results that cannot be generalized as TOEFL test-takers' fundamental problems since the participants in this study were limited and based on specific criteria. Further studies can explore a representative and a larger number of participants involved from several universities. Besides, the study results reveal the most challenging section in the test that is still too general. Thus, further research needs to analyze and study which listening section skills that fail the TOEFL test-takers.

REFERENCES

Bai, Y. (2020). The relationship of test-takers' learning motivation, attitudes towards the actual test use, and test performance of the College English Test in China. *Language Testing in Asia*, 10(1), 1–18.

Barnes, M. (2016). The washback of the TOEFL iBT in Vietnam. *Australian Journal of Teacher Education*, 41(7), 158–174.

Brindley, G., & Slatyer, H. (2002). Exploring task difficulty in ESL listening assessment. *Language Testing*, 19(4), 369–394.

Chang, A. C., Wu, B. W. P., & Pang, J. C. (2013). Second language listening difficulties perceived by low-level learners. *Perceptual and motor skills*, 116(2), 415–434.

Cho, Y., & Blood, I. A. (2020). An analysis of TOEFL® Primary™ repeaters: How much score change occurs? *Language Testing*, 37(4), 503–522.

Chou, M. H. (2015). Impacts of the Test of English Listening Comprehension on students' English learning expectations in Taiwan. *Language, Culture, and Curriculum*, 28(2), 191–208.

Fan, J. (2014). Chinese test-takers' attitudes towards the Versant English Test: a mixed-methods approach. *Language Testing in Asia*, 4(1), 6.

Galikyan, I., Madyarov, I., & Gasparyan, R. (2019). Student Test Takers' and Teachers' Perceptions of the TOEFL Junior® Standard Test. *ETS Research Report Series*, 2019(1), 1–15.

Ginther, A., & Elder, C. (2014). A comparative investigation into understandings and uses of the TOEFL iBT® test, the International English Language Testing Service (Academic) test, and the Pearson Test of English for graduate admissions in the United States and Australia: A case study of two university contexts. *ETS Research Report Series*, 2014(2), 1–39.

Hung, S. T. A., & Huang, H. T. D. (2019). Standardized proficiency tests in a campus-wide English curriculum: a washback study. *Language Testing in Asia*, 9(1), 1–17.

Knoch, U., Huisman, A., Elder, C., Kong, X., & McKenna, A. (2020). Drawing on repeat test-takers to study test preparation practices and their links to score gains. *Language Testing*, 37(4), 550–572.

Koizumi, R., & Nakamura, K. (2016). Factor structure of the Test of English for Academic Purposes (TEAP®) test in relation to the TOEFL iBT® test. *Language testing in Asia*, 6(1), 3.

Kostin, I. (2004). Exploring item characteristics that are related to the difficulty of TOEFL dialogue items. *ETS Research Report Series*, 2004(1), i–59.

Liu, O. L. (2014). Investigating the relationship between test preparation and TOEFL iBT® performance. *ETS Research Report Series*, 2014(2), 1–13.

Mickan, P., & Motteram, J. (2008). An ethnographic study of classroom instruction in an IELTS preparation program. *International English Language Testing System (IELTS) Research Reports 2008: Volume 8*, 1.

Namaziandost, E., Hafezian, M., & Shafiee, S. (2018). Exploring the association among working memory, anxiety, and Iranian EFL learners' listening comprehension. *Asian-Pacific Journal of Second and Foreign Language Education*, 3(1), 1–17.

Nassaji, H. (2015). Qualitative and Descriptive Research: Data Type Versus Data Analysis, *Language Teaching Research*, 2015, Vol. 19(2) 129–132

Ockey, G. J., & Gokturk, N. (2019). Standardized Language Proficiency Tests in Higher Education. Second Handbook of English Language Teaching, 377–393.

O'Dwyer, J., Kantarcıoğlu, E., & Thomas, C. (2018). An Investigation of the Predictive Validity of the TOEFL iBT® Test at an English-Medium University in Turkey. *ETS Research Report Series*, 2018(1), 1–13.
Pan, Y. C. (2016). Learners' Perspectives of Factors Influencing Gains in Standardized English Test Scores. *TEFLIN Journal*, 27 (1), January 63–81.
Pearson, W. S. (2019). 'Remark or retake'? A study of candidate performance in IELTS and perceptions towards test failure. *Language Testing in Asia*, 9(1), 17.
Révész, A., & Brunfaut, T. (2013). Text characteristics of task input and difficulty in second language listening comprehension. *Studies in Second Language Acquisition*, 35(1), 31–65.
Tsai, Y., & Tsou, C. H. (2009). A standardized English language proficiency test as the graduation benchmark: Student perspectives on its application in higher education. *Assessment in Education: Principles, Policy & Practice*, 16(3), 319–330.
Wu, J., & Lee, M. C. L. (2017). The relationships between test performance and students' perceptions of learning motivation, test value, and test anxiety in the context of the English benchmark requirement for graduation in Taiwan's universities. *Language Testing in Asia*, 7(1), 9.
Yin, R. K. (2017). *Case study research and applications: Design and methods*. Sage publications.

Post Pandemic L2 Pedagogy – Adi Putra & Arifah Drajati (Eds)
© 2021 Taylor & Francis Group, London, ISBN 978-1-032-05807-8

Students' perceived use of metacognitive strategies in reading shifting multimodal text modes

Theresia Manalu & Yanty Wirza
Universitas Pendidikan Indonesia

ABSTRACT: The current paper explores the profile of metacognitive strategies of high achieving students when reading visual, audio, and linguistic text modes in a public senior high school in Indonesia based on a case study design. The data are gathered from 68 EFL students using reading comprehension tests, MARSI questionnaire (Mokhtari & Reichard 2002), and retrospective think-aloud session. The questionnaires are analyzed inferentially and descriptively. The verbal data are analyzed based on the constructivist theoretical perspective of students' perceptions, beliefs, and experiences (Koro-Ljungberg et al. 2013). Findings show the absence of low awareness of global, problem solving, and support reading strategies and significant metacognitive reading strategies difference among three different modes shifted. The purposes of the strategies employed are fully discussed. The research concludes the learning styles and the text authenticity are impactful to the metacognitive strategies awareness. The finding warrants the need for text authenticity to stimulate students' metacognitive awareness.

Keywords: Audio representation, authentic texts, linguistic representation, metacognitive reading strategies, visual representation

1 INTRODUCTION

The provocation of reading comprehension as one of the macro skills to be an essential array is typically spreading in the Indonesian EFL teaching and learning sphere. Nonetheless, according to PISA, the Indonesian students' proficiency in reading comprehension is low, as one of the most popular literacy assessments. The inadequacy of reading different kinds of texts is considered as one of Indonesian students' initial drawbacks in literacy (Harsiati 2018). The multiple modes of integrating the reading texts in PISA demonstrate the more significant role of multimodal texts in combining multi-semiotic resources in text composition. This manifests the text immersed in reading instructions to expose today's real communicative context in which teachers should use to carry out.

In response, the metacognitive strategy is considered a powerful strategy to be taught in the classroom. Several studies affirmed a close relationship between the positive reading output and their level of metacognitive reading strategies activation of successful students (Rastegar et al. 2017; Wang 2016). Mohamed et al. (2006) investigated good Malaysian readers on their metacognitive reading awareness and use, issuing the essential meaning of reading strategies and the engagement of students' critical understanding. The results confirmed that good learners are aware of their own metacognitive strategies and frequently used those to comprehend texts. However, this revelation is still debatable due to the inconsistent reading strategies used and the level of students' proficiency (Meniado 2016; Pammu et al. 2014).

These empirical studies reveal considerable insights into the current research. No study has investigated the metacognitive reading strategies of successful students concerning multimodal text modes and their purposes in a particular situation. To bridge this gap, the present study tries to

DOI 10.1201/9781003199267-27

answer the problematic issues captured in case students' metacognitive reading strategies towards multimodal texts.

2 LITERATURE REVIEW

As mentioned earlier, the initial theory of multimodal text is rooted in the theory of multiliteracies, which indicates linguistic plurality and multimodal modes of linguistic expression and representation (Cope & Kalantzis 1996). These multiple combinations of forms of literacies focus on meaning-making processes rather than language production defined as multimodal texts in the form of both printed and digital texts (Bloch 2018; Kress 2003). Consequently, considerable attention has been evoked to understand the meaning-making process of proficient students while reading a text in the concept of strategies they use, their reasons, and the circumstances they activate those strategies (Anderson 2003; Mokhtari & Reichard 2002). Accordingly, metacognitive reading strategy refers to implementing metacognitive awareness in relating those two essential parts of learning, which are cognition and instruction in terms of reading comprehension skills to understand meaningful information and ideas.

Accordingly, studies on reading comprehension tasks and reading strategies are frequently investigated. Chevalier et al. (2017) examined several metacognitive issues regarding the learning strategies, metacognitive and behavioural studies, and the possibility of GPA and metacognitive strategies relationship in both of student groups. They found that students with a history of reading difficulties are less likely to use metacognitive strategies in the meaning making process than students with no reading problems. As a result, Chen and Chen (2015) argued that metacognition helps students to be attentive and successful on meaning of what they have read before, during, and after reading. The use of metacognitive strategies helps students be more thoughtful to think before, during, and after reading. In this case, students will not be aware of what they are reading which getting impact to the activation of their own strategy. Moreover, readers will not activate their own strategy if they don't know the reason and value they read. In regards, students are encouraged to activate various comprehension strategies covering how, when, and why they use it.

A number of research have also portrayed the essential information of students' metacognitive strategies variance on practices and experiences in reading comprehension. In a recent example, Shang (2017)) explored the metacognitive strategies, hypermedia annotations use, and reading comprehension in facilitating EFL students' understanding in Taiwan based on students' proficiency level. He tried to examine whether Taiwanese students' act on reading strategies use hypermedia annotations, and reading comprehension gives different evidence or not to the proficiency levels. The results then revealed that there is no significant difference among those three variables.

To sum up, there has been much research on typically relating students' variances and students' awareness of reading strategies towards various texts. However, still fewer studies have examined students' metacognition in diverse texts composition. Thus, this study confirmed the questions: what kinds of metacognitive reading strategies are frequently activated by high achieving students in reading shifting multimodal text modes, and to what extent they activate those strategies in each shifting multimodal text mode.

3 RESEARCH METHOD

This current study took place at a public exemplary boarding school in Indonesia. The chosen school has been exposing multimodal texts in its English reading materials. The participants are 68 students of grade XI in the academic year 2020 who are then purposely and stratifically selected based on the teacher's judgment, entrance exam, and reading comprehension test scores. Ten students then stratifically chosen, representing high scores cluster to be continuously checked for their answers on MARSI questionnaires of each text mode (Mokhtari & Reichard 2002). Two final selected students (one male and one female) then volunteered in the think-aloud session.

The three-week multimodal-based reading comprehension tests are emphasized and infused in visual, audio, and linguistic text composition. The reading comprehension tests are originally adapted and constructed from several resources that are then confirmed to the local English teacher to suit their teaching syllabus. Thus, *Dr. Martin Luther King, Jr* is chosen as the supplementary topic. The students' empirical and behavioral evidence and cognitive processes are collected. SPSS 26.0 for Windows is used to analyze the results of the MARSI Questionnaire statistically. The think-aloud session is conducted retrospectively through a virtual platform. In doing so, the participant account is transcribed verbatim. The verbal data then are analyzed based on the constructivist perspective of students' perceptions and their beliefs and experiences from individual actions and constructed explanations by Koro-Ljungberg et al. (2013).

4 RESULTS AND DISCUSSION

This section answers the research questions of this study about the tendency to use metacognitive reading strategies by high achieving students and their purposes of activated the current strategy. The study results consist of qualitative and quantitative data to show the activated metacognitive reading strategies of the selected participants and the purpose they activated those strategies in comprehending the three multimodal texts shifted. The complete and detailed results of this study and its analysis are presented in the following finding and discussion.

4.1 *Metacognitive reading strategies frequently use*

4.1.1 *Visual representation mode*

The representational function in visual text mode is used to symbolize or represent the idea in which authors' purpose can be communicated by image infusion. The result shows problem-solving strategies are the most common strategies used by high achieving students when reading visual text mode from the average MARSI questionnaire scores for visual text mode. It indicates that high achieving students have tried to skillfully navigate their action plans through the text when the text is difficult to understand during the meaning-making process in reading visual text mode. Those overall results presented in Table 1 then confirmed that in the practical use of strategies in the reading text in visual text mode, high achieving students cohesively activate their strategies in frequent and effective ways appropriate to the visual text mode.

Table 1. The overall analysis of reading strategies based on dimensions (N = 10).

Reading Strategies Subscale	Overall Mean (M)	Overall Std. Deviation (SD)
Global Reading Strategies	3.65	0.99
Problem Solving Strategies	**3.9**	**0.74**
Support Reading Strategies	3.33	1.02

4.2 *Audio representation mode*

Audio or aural representation represents the audio semiotic mode of multimodal design in the form of sound. The average score of audio text mode is found different results from the previous text mode. The global reading strategy is the most frequent use of metacognitive reading strategies than the other two metacognitive reading strategies. This phenomenon indicates that successful students exhibit their metacognitive reading strategies to the global analysis of the text they are reading in reading audio text mode. The most significant sub-category from global reading strategies constructed by high achieving students with the highest average score is concerned with the purpose

Table 2. The overall analysis of reading strategies based on dimensions (N = 10).

Reading Strategies Subscale	Overall Mean (M)	Overall Std. Deviation (SD)
Global Reading Strategies	**4.12**	**0.89**
Problem Solving Strategies	3.99	0.77
Support Reading Strategies	3.92	0.91

in mind when they are reading. It indicates that high achieving students set their specific reason before reading audio text mode to avoid overload purposes, which can then be tapped to activate their prior knowledge.

4.2.1 *Linguistic representation mode*

Linguistic representation in text plays a role in representing meaning through writing and reading in handwriting, a printed page, and a screen page. A similar result is found to the result of visual text mode in which problem-solving strategies take the highest overall mean scores. The most significant sub-category of problem-solving strategies constructed refers to paying close attention to what they read as the text gets hard to read. It seems high achieving students struggle to check and recheck their own understanding of conflicting information for their better experience when reading linguistic text mode.

Table 3. The overall analysis of reading strategies based on dimensions (N = 10).

Reading Strategies Subscale	Overall Mean (M)	Overall Std. Deviation (SD)
Global Reading Strategies	4.29	0.70
Problem Solving Strategies	**4.31**	**0.73**
Support Reading Strategies	3.9	0.98

4.3 *The extent of high achieving students activate their metacognitive reading strategies*

4.3.1 *Global reading strategies*

There are five dominant notable strategies of global reading strategies from two volunteered students during the think-aloud session, presented in Table 4. Both students spontaneously and explicitly explained how they started reading. It can be assumed that both students have the same purpose but different reasons. Thus, finding shows that textual features, composed in the text, help students visualize and highlight the text's particular context, especially in reading linguistic text mode. Student A described what she knew in detail how she envisioned the text she read. She gave her personal comment by providing a conclusion that the text is composed of the true story due to the title, pictures, and color she highlighted.

Another personal comment noted, defined in Student A's personal comparison of the actual situation nowadays from the camera filters she proposed and the factual reality. At the same time, student B focused on the subtitle in the visual text mode to get initial information. One of them seems to already make a one-page summary in the Indonesian language towards visual text modes by collecting all information in her mind and personal notes. For that reason, Mokhtari (2018) argued that mediated reading purpose to prior knowledge activation in reading comprehension can keep the students from becoming overwhelmed by finding specific reasons for their reading purpose.

Table 4. The repertoire of global reading strategies.

Metacognitive reading strategies	Students' reasons
5. Setting a purpose for reading	a. To understand the task, she would do it by firstly reading the instructions. b. To know what is being assigned and to understand what he should do with the task.
6. Activating prior knowledge	a. To elicit and to build the initial knowledge, they need to predict the upcoming content or information. b. To guess certain textual features
7. Predicting what text is about	a. To observe the topic and to predict the next information b. To give a hypothesis of what comes next in the text
8. Using colors, pictures, and other textual features	a. To visualize the text to be read realized from pictures, title, and color b. To highlight the specific phenomenon of the context of a text
9. Making a summary	a. To draw and to demonstrate personal conclusions in mind to a written form.

4.3.2 *Problem solving strategies*

The findings of verbal data collection in problem-solving strategies support the statistical results revealed previously as the dominant strategies frequently used by high achieving students in the current study. All problem-solving strategies are activated actively in three text modes assigned, noticed by the students' behavioral and gestural activities. These activities are also considered when both students seem to read the text and the questions loudly. It is also found that those two students replayed the video and the record and tried to pay attention to the several scenes. Pausing, rewinding, and paying close attention to a particular sentence and scene are dominant strategies found in visual and audio text mode. Problem-solving strategies are typically used when problems arise. In other words, this strategy are used when it is difficult to understand the particular part of a text. The reader rereads the text to predict the meaning of unfamiliar words (Mokhtari & Reichard 2002). Similar findings also noted from Yüksel and Yüksel (2012) indicated that the students are used to using reading strategies when reading, so they used to be aware of reading strategies. He also declares that problem-solving strategies are the most frequent use and the highest awareness of Turkey's students.Similarly, strategy awareness patterns while reading academic texts would be the same. However, different groups of students have been considered in significantly different contexts of proficiency and sociocultural environments. These kinds of strategies are essential to help students comprehend a text and achieve a higher level of thinking in a particular task (Goh 2008).

Table 5. The repertoire of problem-solving strategies.

Metacognitive reading strategies	Students' reasons
1. Reading slowly and carefully	To focus on getting the key of ideas
2. Pausing to reflect on reading	To gather information before and after a particular sentence or scene
3. Paying close attention to reading	To process the knowledge got before and after reading particular information
4. Rereading	To make sure that the information got have been appropriate to the predictive information
5. Adjusting the reading rate	To pay attention to important information in the text
6. Guessing the meaning of unknown words	To continue collecting information

4.3.3 *Support reading strategies*

The information on support reading strategies according to the think-aloud session is wholly presented in Table 6. It provides the activity of paraphrasing the text information that both students use in the whole process of the think-aloud session, especially in reading audio and linguistic text modes by rephrasing sentences or speech in their own words. The different points of view noted towards student A and student B processes of paraphrasing. If student A strictly stated her opinion based on the text as the third person, Student B freely involved himself in the text. This brings him closer to the story and creates the ultimate intimacy with the story he already read.

Both students also employed the same action of using Google to translate difficult words into the Indonesian language. After getting the meaning, they will write it down as their personal notes while forcing themselves to link unknown words and continually interpret the text's information. From the evidence provoked earlier, maximizing the familiar aids is the students' choice to force themselves understanding a word's meaning to get a deeper understanding.

In the current study, both Student A and Student B worked on their notes while reading a text, which dominantly infused audio design. The evidence of their thinking process then is proven from their behavioral evidence. From student A's notes, it shows that she realized the text organization by giving marks "opening speech" to the opening paragraph and "closing speech" to the end of the text. In contrast, student B focused more on the contextual information of the text. Similar findings are also revealed from Shang (2017), who found that paraphrasing, using supporting aids, and highlighting are much more frequently used than other strategies within supporting reading strategies. This strategy's activation is as the supporting action, conveniently done by high achieving students to connect words that are emphasized by their self-questions to get a deeper determination of their prediction.

Table 6. The repertoire of support reading strategies.

Metacognitive reading strategies	Students' reasons
1. Paraphrasing text information	a. To clarify and to simplify the complex information or ide in the text b. To get simple meaning from the text
2. Using Google to translate difficult words (supporting aid)	a. To link between unknown words b. To continue reading
3. Taking annotations, underlining, and circling text information	a. To make a sign on the text organization b. To focus on particular words heard and contextual information

5 CONCLUSION

This current study tries to draw students' metacognitive reading strategies to towards multimodal text composition. To give wider portrait of this phenomenon, this study tries to conceptualize the strategy used in case how students of high achievers activate their own reading strategies. This study confirmed the questions on how and what the phenomenon of metacognitive reading strategies are revealed.

In conclusion, the text modes' heterogeneity makes differences in the reading strategies of high achieving students when activating their metacognitive reading strategies on thinking during reading activity. The spontaneous activation of metacognitive reading strategies helps students to be more aware of text features and modes navigation in negotiating new meaning resulting from the interaction and integration of modes. It makes no significant difference among the three different modes assigned to the students. It is evident from the findings that students with high performance perform a medium and high level of awareness on almost all MARSI subscales. This study implies the critical issues of selecting reading texts or materials that may capture students' learning styles and authentic reading materials to encourage students' meaning-making process.

Further research is required to investigate why particular strategies are used at various levels of reading performance. Among other things, it is recommended to investigate the students learning style because it may further demonstrate which strategies are implemented. The interaction between metacognitive reading strategies and learning styles in EFL students can perhaps be also examined in future research. There is a clear need to investigate the role of teaching strategies empirically and study their effect on students' reading. It is more highly recommended to examine the concept of strategy in particular circumstances rather than merely knowing what strategies to use to shed a new landscape on metacognitive reading strategies when students are involved in reading activities.

REFERENCES

Anderson, N. (2003). Scrolling, clicking, and reading English: Online reading strategies in a second/foreign language. *The Reading Matrix, 3*(3).

Bloch, J. (2018). Digital storytelling in the multilingual academic writing classroom: Expanding the possibilities. *Dialogues: An Interdisciplinary Journal of English Language Teaching and Research, 2*(1), 96–110.

Chen, K. T. C., & Chen, S. C. L. (2015). The use of EFL reading strategies among high school students in Taiwan. *The Reading Matrix: An International Online Journal, 15*(2), 156–166.

Chevalier, T. M., Parrila, R., Ritchie, K. C., & Deacon, S. H. (2017). The role of metacognitive reading strategies, metacognitive study and learning strategies, and behavioral study and learning strategies in predicting academic success in students with and without a history of reading difficulties. *Journal of Learning Disabilities, 50*(1), 34–48.

Cope, B., & Kalantzis, M. (1996). New Literacies, New Learning. *Multiliteracies*, 1–30.

Harsiati, T. (2018). Karakteristik soal literasi membaca pada program PISA. *Litera, 17*(1), 90–106.

Koro-Ljungberg, M., Douglas, E. P., Therriault, D., Malcolm, Z., & McNeill, N. (2013). Reconceptualizing and decentering think-aloud methodology in qualitative research. *Qualitative Research, 13*(6), 735–753.

Kress, G. (2003). Literacy in the New Media Age (review). In *Routledge*. Taylor & Francis.

Meniado, J. C. (2016). Metacognitive reading strategies, motivation, and reading comprehension performance of Saudi EFL students. *English Language Teaching, 9*(3), 117.

Mohamed, A. R., Chew, J., & Kabilan, M. K. (2006). Metacognitive reading strategies of good Malaysian Chinese learners. *Malaysian Journal of ELT Research, 2*(March), 21–41.

Mokhtari, K. (2018). Prior knowledge fuels the deployment of reading comprehension strategies. *The TESOL Encyclopedia of English Language Teaching, 1984*, 1–10.

Mokhtari, K., & Reichard, C. A. (2002). Assessing students' metacognitive awareness of reading strategies. *Journal of Educational Psychology, 94*(2), 249–259.

Pammu, A., Amir, Z., & Maasum, T.N.R.M.T. (2014). Metacognitive reading strategies of less proficient tertiary learners: A case study of EFL learners at a public university in Makassar, Indonesia. *Procedia – Social and Behavioral Sciences, 118*, 357–364.

Rastegar, M., Mehrabi Kermani, E., & Khabir, M. (2017). The relationship between metacognitive reading strategies use and reading comprehension achievement of EFL learners. *Open Journal of Modern Linguistics, 07*(02), 65–74.

Shang, H. F. (2017). Exploring metacognitive strategies and hypermedia annotations on foreign language reading. *Interactive Learning Environments, 25*(5), 610–623.

Wang, Y. H. (2016). Reading strategy use and comprehension performance of more successful and less successful readers: A think-aloud study. *Kuram ve Uygulamada Egitim Bilimleri, 16*(5), 1789–1813.

Post Pandemic L2 Pedagogy – Adi Putra & Arifah Drajati (Eds)
© 2021 Taylor & Francis Group, London, ISBN 978-1-032-05807-8

Author index